The Science of Electric Vehicles

The Science of Electric Vehicles: Concepts and Applications presents the basic electrical principles, physics, chemistry, use of rare earth elements (REEs), batteries, charging, and operation of motor controllers of EVs. In addition to the general concepts, the book examines the policies and economics pertinent to the move from hydrocarbon power to electric-powered vehicles. It covers the history and development of electric vehicles as well as the science and engineering behind them.

Features:

- Presents the basic electrical principles, physics, and chemistry involved in the manufacture of electric vehicles.
- Discusses various battery types, energy efficiency, storage, and more.

The Science of Electric Vehicles

Concepts and Applications

Frank R. Spellman

CRC Press is an imprint of the
Taylor & Francis Group, an **informa** business

Designed cover image: Shutterstock | Gorodenkoff

First edition published 2023
by CRC Press
6000 Broken Sound Parkway NW, Suite 300, Boca Raton, FL 33487-2742

and by CRC Press
4 Park Square, Milton Park, Abingdon, Oxon, OX14 4RN

CRC Press is an imprint of Taylor & Francis Group, LLC

Library of Congress Cataloging-in-Publication Data
Names: Spellman, Frank R., author.
Title: The science of electric vehicles : concepts and applications / Frank R. Spellman.
Description: First edition. | Boca Raton : CRC Press, 2023. | Includes bibliographical references and index. |
Identifiers: LCCN 2022044179 (print) | LCCN 2022044180 (ebook) | ISBN 9781032366289 (hardback) | ISBN 9781032366296 (paperback) | ISBN 9781003332992 (ebook)
Subjects: LCSH: Electric vehicles.
Classification: LCC TL220 .S647 2023 (print) | LCC TL220 (ebook) | DDC 629.22/93–dc23/eng/20221206
LC record available at https://lccn.loc.gov/2022044179
LC ebook record available at https://lccn.loc.gov/2022044180

ISBN: 978-1-032-36628-9 (hbk)
ISBN: 978-1-032-36629-6 (pbk)
ISBN: 978-1-003-33299-2 (ebk)

DOI: 10.1201/9781003332992

Typeset in Times
by codeMantra

Contents

List of Acronyms and Abbreviations

3D: three-dimensional
AC: alternating current
ACB: active current balancing
AEV: all-electric vehicle
AGD: active gate driver
Al: aluminum
Alnico: (Al-Ni-Co-Fe, family of iron alloys)
AWG: American wire gauge
bemf: back electromotive force
BREM: beyond rare earth magnets
CAN: controller area network
CBC: current balancing controller
CDS: combined driving schedule
CFD: computational fluid dynamics
CGD: convention gate driver
CMOS: complementary metal-oxide semiconductor
CMR: common mode reaction
CT: current transducer
Cu: copper
CVD: chemical vapor deposition
DBC: direct bond copper
DC: direct current
DCR: dc resistance
DCT: differential current transformer
DOE: US Department of Energy
DSP: digital signal processing/processor
Dy: dysprosium
eGaN: enhancement mode gallium nitride
EM: electric machine
EMC: epoxy molding compound
emf: electromotive force
EMI: electromagnetic interference
EPA: Environmental Protection Agency
EPC: efficient power conversion
ESR: equivalent series resistance
EV: electric vehicle
FEA: finite element analysis
FET: field-effect transistor
FUL: fault under load

GaN: gallium nitride
GIR: gate impedance regulation
GPM: gallon per minute
GVC: gate voltage control
HEMT: high electron mobility transfer
HEV: hybrid electric vehicle
HIL: hardware in loop
HSF: hard switching unit
HSG: hybrid starter-generator
HV: high voltage
HWFET: Highway Fuel Economy Test
IC: integrated circuit
IGBT: insulated gate bipolar transistor
IM: induction motor/machine
IPM: interior permanent magnet
IR: insulation resistance
JBS: junction barrier Schottky
JFE: JFE Steel Corporation
JFET: junction field-effect transistor
K: thermal conductivity
K: degrees Kelvin
M/G: motor generator
MFP: multiple isolated flux path
MOSFET: metal-oxide semiconductor field-effect transistor
Nd: neodymium
NREL: National Renewable Energy Laboratory (DOE)
OBC: on-board charger
OD: outer diameter
OEM: original equipment manufacturer
ORNL: Oak Ridge National Laboratory
PCB: printed circuit board
PCU: power converter unit
PD: power density (peak)
PE: power electronics
PEV: plug-in electric vehicle
PF: power factor
PFC: power factor correction
PM: permanent magnet
PSAT: Powertrain Systems Analysis Toolkit
PSIM: Powersim (circuit simulation software)
PWM: pulse width modulated/modulation
PwrSoC: power supply on chip
R&D: research and development
RC: resistor-capacitor
RE: rare earth
RESS: regenerative energy storage system

RL: resistor-inductor
rms: root mean square
SBD: Schottky barrier diode
SCC: switched capacitor converter
Si: silicon
SiC: silicon carbide
SJT: super junction transistor
SOC: state of charge
SOI: silicon-on-insulator
SP: specific power
SPM: surface permanent magnet
SRM: switched reluctance motor
SSCB: solid-state circuit breaker
TC: thermal conductivity
TDS: traction drive system
THD: total harmonic distortion
UC: ultracapacitor (aka supercapacitor)
VAC: voltage AC
Vce: voltage across collector and emitter
VDC: voltage DC
VGD: variable gate delay
VSI: voltage source inverter
WBG: wide bandgap
ZS: zero sequence

Preface

The Science of Electric Vehicles: Concepts and Applications is the eighth volume in the acclaimed series that includes *The Science of Rare Earth Elements: Concepts and Applications*, *The Science of Water*, *The Science of Air*, *The Science of Environmental Pollution*, *The Science of Renewable Energy*, *The Science of Waste*, and *The Science of Wind Power*, all of which bring this highly successful series fully into the 21st century. *The Science of Electric Vehicles: Concepts and Applications* continues the series mantra based on good science and not feel-good science. It also continues to be presented in the author's trademark conversational style.

This practical, direct book presents technologies and techniques, as well as construction of and operation of today's electric vehicles. This book is designed to be used as an information source for the general reader, or for a course in *Electric Vehicle Science/Engineering*—and, as stated, is presented in the author's typical plain English approach. In this book, you will find basic electrical principles, physics, chemistry, use of rare earth elements (REEs), batteries, charging, and operation of motor controllers—again, all is presented in understandable down-to-earth prose; no convoluted verbiage. Moreover, beyond these concepts, many applications are highlighted including policies and economics pertinent to the move from hydrocarbon power to electric-powered vehicles. Eventually moving from fossil fuel power to electric power will be critical to sustaining our modern way of life. This book focuses on answering the following questions: What are electric vehicles? What is the history of electric vehicles and their use? What is the science and engineering of electric vehicles? And also, what does the future hold for the usage of electric vehicles? Moreover, this book asks the same basic and pertinent question related to the topic of discussion: Why should we care about electric vehicles?

This last question (and the answer provided in the text) is or should be of particular interest to those who are advocates for the use of renewable energy sources and a pollution-free environment. Concern for the environment and the impact of environmental pollution has brought about the trend (and the need) to shift from the use and reliance on hydrocarbons to energy power sources that are pollution neutral or near pollution neutral and renewable. We are beginning to realize that we are responsible for much of the environmental degradation of the past and present—all of which is readily apparent today. Moreover, the impact of 200 years of industrialization and surging population growth has far exceeded the future supply of hydrocarbon power sources. So, the implementation of renewable energy sources is surging, and along with it, there is a corresponding surge in utilization of electric vehicles as our primary means of transportation.

Why a text on the science of electric vehicles? Simply put, studying physics, electricity, motion, materials, metals, products, and so forth without including the inherent science connection is analogous to attempting to reach an unknown, unfamiliar location without being able to read a map, written directions, or digital device.

Many of us have come to realize that a price is paid (sometimes a high price) for what is called "the good life." Our consumption and use of the world's resources make all of us at least partially responsible for pushing the need to prevent the pollution of our environment due to our use of conventional energy sources such as gasoline and coal. Pollution and its ramifications are one of the inevitable products of the good life we all strive to attain, but obviously pollution is not something caused by any single individual, nor can one individual totally prevent or correct the situation. The common refrain we hear today is to reduce pollution and its harmful effects; everyone must band together as an informed, knowledgeable group and pressure the elected decision makers to manage the problem now and in the future. At this moment in time, there is an ongoing push to substitute fossil fuels with renewable energy sources—this is where the shift to renewable energy sources comes into play and where the need to use wind and solar technology, energy storage applications, and electric vehicles is vital.

Throughout this text, common-sense approaches and practical examples have been presented. Again, because this is a science text, I have adhered to scientific principles, models, and observations. But you need not be a scientist to understand the principles and concepts presented. What is needed is an open mind, a love for the challenge of wading through all the information, an ability to decipher problems, and the patience to answer the questions relevant to each topic presented. The text follows a pattern that is nontraditional; that is, the paradigm used here is based on real-world experience, not on theoretical gobbledygook. Real-life situations are woven throughout the fabric of this text and presented in straightforward, plain English to give the facts, knowledge, and information to enable understanding needed to make informed decisions.

Environmental issues are attracting ever-increasing attention at all levels. The problems associated with these issues are compounded and made more difficult by the sheer number of factors involved in managing any phase of any problem. Because the issues affect so many areas of society, the dilemma makes us hunt for strategies that solve the problems for all, while maintaining a safe environment without excessive regulation and cost—Gordian knots that defy easy solutions.

The preceding statement goes to the heart of why this text is needed. Presently, only a limited number of individuals have sufficient background in the science of electric vehicles and their concepts and applications in the world of industrial and practical functions, purposes, and uses to make informed decisions on 21st-century product production, usage, and associated environmental issues.

Finally, *The Science of Electric Vehicles* is designed to reach a wide range of practitioners and students and also to provide a basic handbook or reference for technicians, and those in other industries such as electric vehicle production and maintenance.

The bottom line: Critical to solving these real-world environmental problems is for all of us to remember that old saying, we should take nothing but pictures, leave nothing but footprints, kill nothing but time, and sustain ourselves with the flow of clean, safe, renewable energy—and truck on under electrical power.

Frank R. Spellman
Norfolk, VA

Author

Frank R. Spellman, PhD, is a retired US Naval Officer with 26 years active duty and also a full-time adjunct assistant professor of environmental health at Old Dominion University, Norfolk, Virginia, and the author of more than 157 books covering topics ranging from a 14-volume homeland security series; several safety, industrial hygiene, and security manuals; and also including concentrated animal feeding operations (CAFOs) to all areas of environmental science and occupational health. Many of his texts are readily available online at Amazon.com and Barnes and Noble.com, and several have been adopted for classroom use at major universities throughout the United States, Canada, Europe, and Russia; two have been translated into Chinese, Japanese, Arabic, and Spanish for overseas markets. Dr. Spellman has been cited in more than 850 publications. He serves as a professional expert witness for three law groups and as an incident/accident investigator and security expert for the US Department of Justice and a northern Virginia law firm. In addition, he consults on homeland security vulnerability assessments for critical infrastructures including water/wastewater facilities nationwide and conducts pre-Occupational Safety and Health Administration (OSHA)/Environmental Protection Agency (EPA) audits throughout the country. Dr. Spellman receives frequent requests to co-author with well-recognized experts in several scientific fields; for example, he is a contributing author of the prestigious text *The Engineering Handbook* (2nd edition, CRC Press). Dr. Spellman lectures on wastewater treatment, water treatment, and homeland security and safety topics throughout the country and teaches water/wastewater operator short courses at Virginia Tech (Blacksburg, Virginia). In 2011–2012, he traced and documented the ancient water distribution system at Machu Pichu, Peru, and surveyed several drinking water resources in Amazonia-Coco, Ecuador. Dr. Spellman also studied and surveyed two separate potable water supplies in the Galapagos Islands; he also studied and researched Darwin's finches while in the Galapagos. He holds a BA in public administration, a BS in business management, an MBA, and an MS and PhD in environmental engineering.

NOTE TO THE READER

A couple of items are of importance. First, when we address fluid, we are including air; this is standard engineering and science-related accepted practice. Second, when we refer to vehicles-on-wheels, we are pinpointing those vehicles (automobiles and trucks) that travel the highways, even though motorcycles and trains are obviously vehicles-on-wheels and are and can be included in various calculations and realistic parameters—but the focus is on vehicles-on-wheels that travel highways, streets, farmland, or wherever.

1 Electric Vehicles (EVs)

INTRODUCTION

In a non-scientific manner, I visited several car dealerships where EVs were being displayed and sold and asked a salesperson why I should buy an EV instead of a standard gas-/diesel-powered vehicle and without a moment's hesitation the salesperson replied:

"EVs are:

- Fun to drive because they are smooth and fast.
- You do not have to worry about producing harmful greenhouse gases that will eventually destroy us all.
- Now, these two models here to our right and left are state-of-the-art products and very popular.
- Also, keep in mind that the experts and their studies show that emissions from burning fossil fuels such as gasoline produce greenhouse gases. Now you two look to me to be concerned about our environment … so, also keep in mind that the EVs produce no smelly fumes or harmful greenhouse gases.
- Another thing, EVs are cool, fun to drive, and very innovative.
- EVs only cost pennies a year to operate compared to thousands of Washingtons (U.S. dollar bills) for gas to operate them guzzlers.
- And, oh, by the way, think about that great tax break you're gonna get by buying one of these beauties—we're talking thousands, folks … yes indeed.
- The bottom line: only the very smart and environmentally conscious are purchasing these great vehicles."

Author's note: The preceding EV sales event(s) actually took place on three different occasions at three different dealerships when the author was the potential customer but actually was conducting non-scientific research and not actually purchasing—just a tire kicker, so to speak.

Now, after listening to several EV sales pitches on why it would be smart to buy a new EV I also need to point out that as a potential customer I asked a question that even the salesperson acknowledged was the most frequently asked question: "How far can I drive before I have to recharge?" Well, I found the standard reply to my inquiry was something along the lines of "do you use a cell phone? And if so, do you charge it up every day? Hey, same thing with your brand-new EV … just charge it up at home … easy … and furthermore the driving range on some of these new EVs is more than 100 km before needing to be recharged … and for most folks that's plenty of range for them."

Well, you probably are wondering what another frequently asked question regarding a new EV might be and I found that it is: "What is the difference between an EV

DOI: 10.1201/9781003332992-1

and a Hybrid?" I found that when I asked this question from someone who really knows Hybrids, the answer usually goes along these lines (I have generalized these answers): "At the present time there are 3 types of electric vehicles: Battery Electric Vehicle (BEV), Plug-in Hybrid electric vehicle (PHEV), and the Hybrid Electric Vehicle (HEV). The point is that there are various models at various prices that satisfy various needs."

And this is what we are told, in one fashion, form, scheme, sales pitch, or another.

One of the questions that I asked, not mentioned to this point, is: "Can you tell me how one of these Electric Vehicles … of any type or brand or model of these EVs actually work?"

Ah! And now we have reached the point of this book … the focus … the message and what EVs are all about. And that starts with electricity … so let's get to it.

To begin this discussion, we need to begin at the beginning. The beginning with electric vehicles is electricity. And we begin this discussion in the following chapters.

2 Electron Flow = Traffic Flow

BACK IN THE GOOD OLD DAYS

Well, back in the good old days (1828–1835) in America, the only traffic flow problems the driver and riders in a horse and buggy rig had were primarily limited to traffic flow in different directions in the central part of town or major city—outside the town or city one could ride with the wind with little traffic of any kind to worry about, so to speak. The horse and/or horse and buggy was it. There were no airplanes, railroads, or automobiles yet, so traffic flow was either by horse or mule or by simply walking to and from wherever you were headed. If you were located in an area where there was some type of traffic flow, you simply had to maneuver your horse and buggy through the traffic. That is, again, if there was traffic at your location.

Fast forward to the 2020s and beyond, a horse and buggy or horse and rider on the main roads today are prime targets of local or state police to get them off the Interstates and main thoroughfares in the cities throughout the United States. Of course, there are exceptions to the horse and buggy operations. In Amish country, for example, horse and buggy are standard means of transport. There are also the celebrations and parades in which paradegoers cherish viewing horses and their riders, and horse-drawn carriages of any configuration (and the obligatory followers steering their wheelbarrows or street sweepers). Paradegoers today enjoy these parades to the fullest … because it is such a novelty to see horses and carriages in the city … and the cleanup crew performing their necessary duties.

HOW TIMES HAVE CHANGED

Wow! How times have changed. Now traffic flow is all about cars, trucks, semi-trucks, delivery vehicles, buses, and very few wheelbarrow pushers with their handy shovels.

Instead, however, we have flow (hopefully) and if you just stop and ascend a vantage point where you can observe the flow of afterwork traffic on an Interstate, you will be amazed and might ask yourself: "Where did they all come from?" And also, "Where are they all going?"

Now if you are a traffic engineer you probably take a different view of traffic flow on the Interstate Highway, or any other thoroughfare and you are not worried about the guy with the shovel and wheelbarrow.

No, horse dung is not the problem. There are other issues with modern-day traffic flow. If you are a traffic engineer you might and probably will look at traffic flow in terms of science—and in this book science makes sense because it portends to be about science.

DOI: 10.1201/9781003332992-2

Anyway, let's look at a bit of engineering science in regard to traffic flow—from the engineering and science point of view.

When attempting to gauge vehicle traffic flow, engineers regularly want to determine vehicle traffic flow and vehicle density.

So, how do the traffic engineers determine traffic flow and vehicle density?

They figure these parameters or factors out in a manner that is quite simple. To determine vehicle traffic flow traffic engineers use the following:

First, we want to know or to determine the vehicle traffic flow and how to express it in our determination. So, we can use q to signify vehicle flow (i.e., vehicle flow—the number of vehicles per hour). Okay, so, if we want to know or determine vehicle traffic flow (we call this q) we need to know the number of vehicles passing in t seconds (we call this n) and, of course, we need to know t (i.e., the time for passing vehicles (s)) and all of this is expressed in Equation (2.1).

$$q = n \times 3{,}600/t \tag{2.1}$$

Okay so let's lay this out for understanding at all levels.

If we make q = vehicle flow (number of vehicles per hour)

And n = number of vehicles passing in t seconds

Then we make t = time for passing vehicles (s)

We develop Equation (2.1), $q = n \times 3{,}660/t$

And to be clearer we present Example 2.1 as follows:

Example 2.1

Problem:

If 2,000 vehicles pass in 2 hours (7,200 s), what is the calculated flow?

Solution:

$$Q = 2{,}000 \cdot 3{,}600/7{,}200\ (s)$$

$$= 1{,}000 \text{ vehicles per hour}$$

Okay, 1,000 vehicles per hour is one of our findings, but we still need to determine vehicle density. In order to determine vehicle density, we can express the equation as shown:

$$K = n \times 5{,}280/l \tag{2.2}$$

where

K = vehicle density (vehicles per mile)

n = number of vehicles occupying a length (l) of the road

l = the length of the road occupied by the vehicles (m)

Well, the preceding information and examples are interesting, informative, and revealing, that is, if we are interested in determining vehicle flow and vehicle density.

However, our interest here and our focus is on and about electric vehicles. So, the preceding information was/is okay, but we need to move on to electric vehicles.

Actually, the preceding information sets us up to move to electric vehicles because the point being made here is that if you want to understand the science of electric vehicles you first must understand the science of electricity—electrical principles, and more (like the kind of equations shown earlier on determining vehicle flow and vehicle density). Traffic engineers and other types of engineers, planners, inventors, and several other professionals today are in the process of shifting gears, so to speak.

Shifting gears in vehicles is either an automatic operation or manual. Well, we are now shifting gears from internal combustion-powered vehicles using hydrocarbon-type fuels to electric-powered vehicles. In doing so (in making this shift), we can go back to the beginning of electric vehicles. In that 1828–1835 timeframe mentioned earlier, and in the age when horse and buggies were the primary mode of transportation, far-thinkers, innovators in Hungary and the Netherlands and the United States, in their far-thinking mode were creating some of the first small-scale electric cars.

Basically, we can say that there was and is a shift from concentrating on vehicle flow to that of electron flow.

Electron flow?

Yes.

What does electron flow have to do with electric vehicles?

Well, actually, electron flow has everything to do with electric vehicles.

How so?

Good question. First, we must acknowledge that electric vehicles are powered by electricity. A no-brainer for sure. But from an engineering and science perspective we must understand what electricity is—what it is all about—how does it, electricity, power vehicles and allow them to join the traffic flow, so to speak.

Secondly, if you were to ask the average person, "what is electricity?" they would probably scratch their head and relate several different answers, with most of them being incorrect or incomplete.

The expected answers will be varied and numerous, with none of them completely accurate.

Why?

Electricity is not easy to explain.

Again, why?

Simply, electricity is difficult to define in any absolute definitive way because we do not know what we do not know about electricity.

However, we know how to produce electricity. We know how to harness electricity. We are well aware of electricity's present applications. For the purpose of this book and this discussion in this book we also know that electricity is a source of power that can be and is used for propelling electric vehicles.

Let's get back to the question of what electricity is. Again, when people are asked this question, after they morph from that confused look on their faces to that straight normal look while trying to answer the question, sooner or later someone will attempt to answer or at least have an idea of what electricity is and will reply that it is that force or power that energizes our lightbulbs, our televisions, our stoves, our hot plates, our electric yard tools, our other power tools, our washing machines and dryers, our refrigerators, our security alarm systems, our iPhones, our computers, our electric razors, our electric vehicles including trains and automobiles, and many other items and appliances of which there are just too many to list.

But let's get back to electricity and what it is. Well, we will try, to the extent that a definitive definition can be made. Truth be told, even the experts in engineering and science fields working with or around or close to electricity in one form or another—that is, in one use or the other—can't tell you exactly what electricity is. Simply, again, we do not know what we do not know about electricity.

However, there is a lot we do know about electricity—well sort of. We use it every day in multitudes of applications; thus, we must know something about it.

And we do know something about it. For purposes of this presentation, we will simply state that electricity is the flow of electrons.

Flow of electrons?

Yes, and we will get to that.

For now, and more importantly, to get even at the margin of the science of electric vehicles, it is imperative that we understand what electricity is and how it works and how it is produced and how it is used in powering vehicles. Therefore, the next chapter of this book presents a discussion of basic electricity so that the reader can understand how electricity is produced and used to propel electrical vehicles (and light up lightbulbs and so on and so forth).

3 Basic Electricity[1]

INTRODUCTION

People living and working in modern societies generally have little difficulty recognizing electrical equipment—electrical equipment is everywhere and (if one pays attention to his or her surroundings) is easy to spot. Despite its great importance in our daily lives, however, few of us probably stop to think what life would be like without electricity. Like air and water, we tend to take electricity for granted. But we use electricity to do many jobs for us every day—from lighting, heating, and cooling our homes to powering our televisions and computers. Then there is the workplace—can we actually perform work without electricity? For example, the typical industrial workplace is outfitted with electrical equipment that

- generates electricity (a generator or emergency generator)
- stores electricity (batteries—batteries store electrical energy and do not create it)
- changes electricity from one level (voltage or current) to another (transformers)
- transports or transmits and distributes electricity throughout the plant site (wiring distribution systems)
- measures electricity (measuring meters/indicators)
- converts electricity into other forms of energy (rotating shafts—mechanical energy, heat energy, light energy, chemical energy, or radio energy)
- protects other electrical equipment (fuses, circuit breakers, or relays)
- operates and controls other electrical equipment (motor controllers)
- converts some condition or occurrence into an electric signal (sensors)
- converts some measured variable to a representative electrical signal (transducers or transmitters)

NATURE OF ELECTRICITY

The word *electricity* is derived from the Greek word "electron" (meaning AMBER). Amber is a translucent (semitransparent) yellowish fossilized mineral resin. The ancient Greeks used the words "electric force" when referring to the mysterious forces of attraction and repulsion exhibited by amber when it was rubbed with a cloth. They did not understand the nature of this force. They could not answer the question, "What is electricity?" This question still remains unanswered. Today, we often attempt to answer this question by describing the effect and not the force. That is, the standard answer given in physics is: electricity is "the force that moves electrons," which is about the same as defining a sail as "the force that moves a sailboat."

[1] Much of the information in this section is adapted from F.R. Spellman *Electricity* (2001) Boca Raton, FL: CRC Press; F.R. Spellman *The Science of Wind Power* (2022) Boca Raton, FL: CRC Press.

DOI: 10.1201/9781003332992-3

At the present time, little more is known than what the ancient Greeks knew about the fundamental nature of electricity, but we have made tremendous strides in harnessing and using it. As with many other unknown (or unexplainable) phenomena, elaborate theories concerning the nature and behavior of electricity have been advanced and have gained wide acceptance because of their apparent truth—and because they work.

Scientists have determined that electricity seems to behave in a constant and predictable manner in given situations or when subjected to given conditions. Faraday, Ohm, Lenz, and Kirchhoff have described the predictable characteristics of electricity and electric current in the form of certain rules. These rules are often referred to as laws. Thus, though electricity itself has never been clearly defined, its predictable nature and easily used energy form has made it one of the most widely used power sources in modern times.

The bottom line: You can "learn" about electricity by learning the rules, or laws, applying to the behavior of electricity and by understanding the methods of producing, controlling, and using it. Thus, this learning about electricity can be accomplished without ever having determined its fundamental identity.

You are probably scratching your head, puzzled.

We understand the main question running through your brain at this exact moment: "This is a section in the text about the physics of electricity and the authors can't even explain what electricity is?"

That is correct; we cannot. The point is no one can definitively define electricity. Electricity is one of those subject areas where the old saying "we don't know what we don't know about it" fits perfectly.

Again, there are a few theories about electricity that have so far stood the test of extensive analysis and much time (relatively speaking, of course). One of the oldest and the most generally accepted theories, concerning electric current flow (or electricity), is known as the *electron theory.*

The electron theory basically states that electricity or current flow is the result of the flow of free electrons in a conductor. Thus, electricity is the flow of free electrons or simply electron flow. And, in this text, this is how we define electricity; that is, again, electricity is the flow of free electrons.

Electrons are extremely tiny particles of matter. To gain an understanding of electrons and exactly what is meant by "electron flow," it is necessary to briefly discuss the structure of matter.

The Structure of Matter

Matter is anything that has mass and occupies space. To study the fundamental structure or composition of any type of matter, it must be reduced to its fundamental components. All matter is made of *molecules*, or combinations of *atoms* (Greek: not able to be divided) that are bound together to produce a given substance, such as salt, glass, or water. For example, if you keep dividing water into smaller and smaller drops, you will eventually arrive at the smallest particle that was still water. That particle is the molecule, which is defined as the *smallest bit of a substance that retains the characteristics of that substance.*

A molecule of water (H_2O) is composed of one atom of oxygen and two atoms of hydrogen. If the molecule of water were further subdivided, there would remain only unrelated atoms of oxygen and hydrogen, and the water would no longer exist as such. Thus, the molecule is the smallest particle to which a substance can be reduced and still be called by the same name. This applies to all substances—solids, liquids, and gases.

Important Point: Molecules are made up of atoms, which are bound together to produce a given substance.

Atoms are composed, in various combinations, of subatomic particles of *electrons, protons*, and *neutrons*. These particles differ in weight (a proton is much heavier than the electron) and charge. We are not concerned with the weights of particles in this text, but the *charge* is extremely important in electricity. The electron is the fundamental negative charge (–) of electricity. Electrons revolve about the nucleus or center of the atom in paths of concentric *orbits*, or shells. The proton is the fundamental positive (+) charge of electricity. Protons are found in the nucleus. The number of protons within the nucleus of any particular atom specifies the atomic number of that atom. For example, the helium atom has 2 protons in its nucleus, so the atomic number is 2. The neutron, which is the fundamental neutral charge of electricity, is also found in the nucleus.

Most of the weight of the atom is in the protons and neutrons of the nucleus. Whirling around the nucleus is one or more negatively charged electrons. Normally, there is one proton for each electron in the entire atom so that the net positive charge of the nucleus is balanced by the net negative charge of the electrons rotating around the nucleus (see Figure 3.1).

Important Point: Most batteries are marked with the symbols + and – or even with the abbreviations POS (positive) and NEG (negative). The concept of a positive or negative polarity and its importance in electricity will become clear later. However, for the moment, you need to remember that an electron has a negative charge and that a proton has a positive charge.

We stated earlier that in an atom the number of protons is usually the same as the number of electrons. This is an important point because this relationship determines the kind of element (the atom is the smallest particle that makes up an element; an element retains its characteristics when subdivided into atoms) in question. Figure 3.2 shows a simplified drawing of several atoms of different materials based on the conception of electrons orbiting about the nucleus. For example, hydrogen has a nucleus consisting of 1 proton, around which rotates 1 electron. The helium

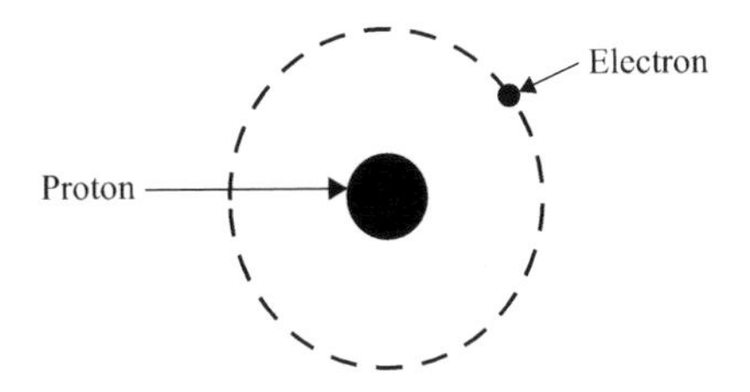

FIGURE 3.1 One proton and one electron = electrically neutral.

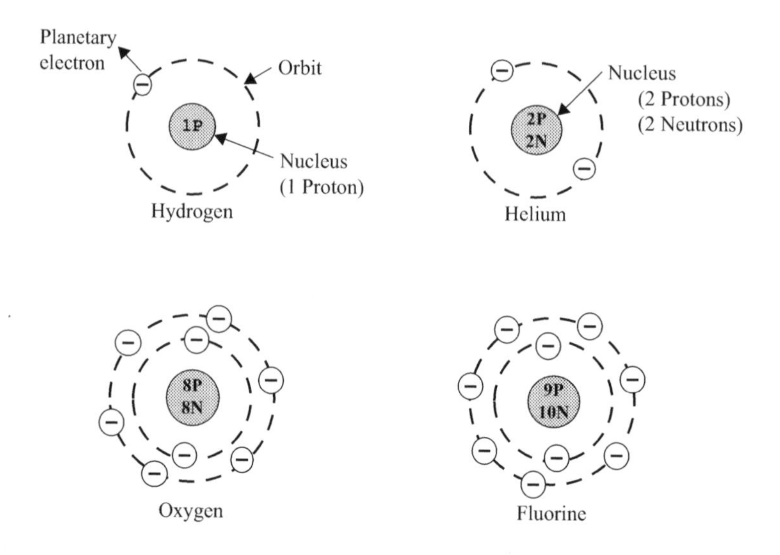

FIGURE 3.2 Atomic structure of elements.

atom has a nucleus containing 2 protons and 2 neutrons with 2 electrons encircling the nucleus. Both of these elements are electrically neutral (or balanced) because each has an equal number of electrons and protons. Since the negative (–) charge of each electron is equal in magnitude to the positive (+) charge of each proton, the two opposite charges cancel.

A balanced (neutral or stable) atom has a certain amount of energy, which is equal to the sum of the energies of its electrons. Electrons, in turn, have different energies called *energy levels*. The energy level of an electron is proportional to its distance from the nucleus. Therefore, the energy levels of electrons in shells farther from the nucleus are higher than that of electrons in shells nearer the nucleus.

When an electric force is applied to a conducting medium, such as copper wire, electrons in the outer orbits of the copper atoms are forced out of orbit (i.e., liberating, or freeing electrons) and impelled along the wire. This electrical force, which forces electrons out of orbit, can be produced in a number of ways, such as: by moving a conductor through a magnetic field; by friction, as when a glass rod is rubbed with cloth (silk); or by chemical action, as in a battery.

When the electrons are forced from their orbits, they are called *free electrons*. Some of the electrons of certain metallic atoms are so loosely bound to the nucleus that they are relatively free to move from atom to atom. These free electrons constitute the flow of an electric current in electrical conductors.

Important Point: When an electric force is applied to a copper wire, free electrons are displaced from the copper atoms and move along the wire, producing electric current as shown in Figure 3.3.

If the internal energy of an atom is raised above its normal state, the atom is said to be *excited*. Excitation may be produced by causing the atoms to collide with particles that are impelled by an electric force as shown in Figure 3.3. In effect what occurs is that energy is transferred from the electric source to the atom. The excess

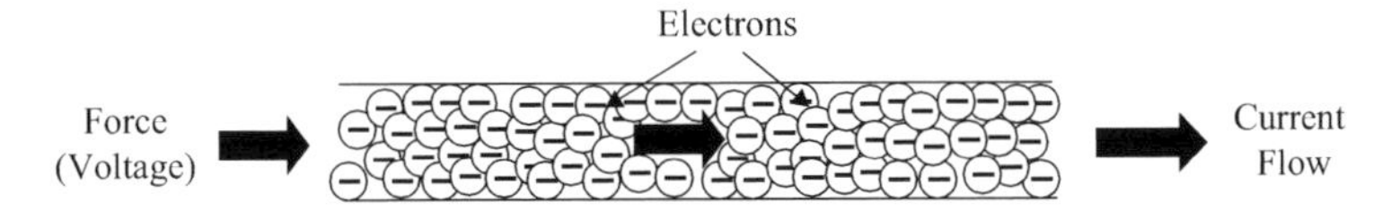

FIGURE 3.3 Electron flow in a copper wire.

energy absorbed by an atom may become sufficient to cause loosely bound outer electrons (as shown in Figure 3.3) to leave the atom against the force that acts to hold them within.

Important Point: An atom that has lost or gained one or more electrons is said to be *ionized*. If the atom loses electrons it becomes positively charged and is referred to as a *positive ion*. Conversely, if the atom gains electrons, it becomes negatively charged and is referred to as a *negative ion*.

CONDUCTORS

Recall that we pointed out earlier that electric current moves easily through some materials but with greater difficulty through others. Three good electrical conductors are copper, silver, and aluminum (generally, we can say that most metals are good conductors). At the present time copper is the material of choice used in electrical conductors. Under special conditions, certain gases are also used as conductors (e.g., neon gas, mercury vapor, and sodium vapor are used in various kinds of lamps).

The function of the wire conductor is to connect a source of applied voltage to a load resistance with a minimum IR voltage drop in the conductor so that most of the applied voltage can produce current in the load resistance. Ideally, a conductor must have a very low resistance (e.g., a typical value for a conductor—copper—is less than 1 Ω per 10 feet).

Because all electrical circuits utilize conductors of one type or another, in this section we discuss the basic features and electrical characteristics of the most common types of conductors.

Moreover, because conductor splices and connections (and insulation of such connections) are also an essential part of any electric circuit, they are also discussed.

UNIT SIZE OF CONDUCTORS

A standard (or unit size) of a conductor has been established to compare the resistance and size of one conductor with another. The unit of linear measurement used is (in regard to diameter of a piece of wire) the **mil** (0.001 of an inch). A convenient unit of wire length used is the **foot**. Thus, the standard unit of size in most cases is the **mil-foot** (i.e., a wire will have unit size if it has diameter of 1 mil and a length of 1 foot). The resistance in ohms of a unit conductor or a given substance is called the **resistivity** (or specific resistance) of the substance.

As a further convenience, **gage** numbers are also used in comparing the diameter of wires. The B and S (Brown and Sharpe) gage was used in the past; now the most commonly used gage is the **American Wire Gage** (AWG).

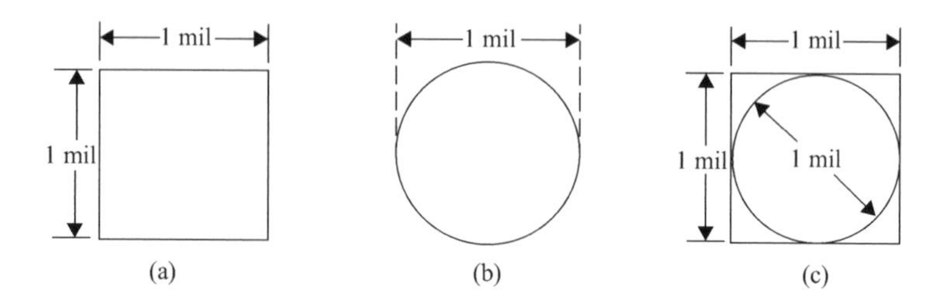

FIGURE 3.4 (a) Square mil; (b) circular mil; and (c) comparison of circular to square mil.

Square Mil

Figure 3.4 shows a square mil. The *square mil* is a convenient unit of cross-sectional area for square or rectangular conductors. As shown in Figure 3.4, a square mil is the area of a square, the sides of which are 1 mil. To obtain the cross-sectional area in square mils of a square conductor, square one side measured in mils. To obtain the cross-sectional area in square mils of a rectangular conductor, multiply the length of one side by that of the other, each length being expressed in mils.

Example 3.1

Problem:

Find the cross-sectional area of a large rectangular conductor 5/8 inch thick and 5 inches wide.

Solution:

The thickness may be expressed in mils as $0.625 \times 1{,}000 = 625$ mils and the width as $5 \times 1{,}000 = 5{,}000$ mils. The cross-sectional area is $625 \times 5{,}000$, or 3,125,000 square mils.

Circular Mil

The *circular mil* is the standard unit of wire cross-sectional area used in most wire tables. To avoid the use of decimals (because most wires used to conduct electricity may be only a small fraction of an inch), it is convenient to express these diameters in mils. For example, the diameter of a wire is expressed as 25 mils instead of 0.025 inch. A circular mil is the area of a circle having a diameter of 1 mil, as shown in Figure 3.4b. The area in circular mils of a round conductor is obtained by squaring the diameter measured in mils. Thus, a wire having a diameter of 25 mils has an area of 25^2, or 625 circular mils. By way of comparison, the basic formula for the area of a circle is

$$A = \pi R^2 \tag{3.1}$$

and in this example the area in square inches is

$$A = \pi R^2 = 3.14(0.0125)^2 = 0.00049 \text{ square inch}$$

If D is the diameter of a wire in mils, the area in square mils can be determined using

$$A = \pi(D/2)^2 \tag{3.2}$$

which translates to

$$= 3.14/4D^2$$

$$= 0.785D^2 \text{ square mils}$$

Thus, a wire 1 mil in diameter has an area of

$$A = 0.785 \times 1^2 = 0.785 \text{ square mils,}$$

which is equivalent to 1 circular mil. The cross-sectional area of a wire in circular mils is therefore determined as

$$A = \frac{0.785D^2}{0.785} = D^2 \text{ circular mils,}$$

where D is the diameter in mils. Therefore, the constant $\pi/4$ is eliminated from the calculation.

Note that in comparing square and round conductors, the circular mil is a smaller unit of area than the square mil, and therefore there are more circular mils than square mils in any given area. The comparison is shown in Figure 3.4c. The area of a circular mil is equal to 0.785 of a square mil.

Important Point: To determine the circular mil area when the square mil area is given, divide the area in square mils by 0.785. Conversely, to determine the square mil area when the circular mil area is given, multiply the area in circular mils by 0.785.

Example 3.2

Problem:

A No. 12 wire has a diameter of 80.81 mils. What is (1) its area in circular mils and (2) its area in square mils?

Solution:

1. $A = D^2 = 80.81^2 = 6{,}530$ circular mils
2. $A = 0.785 \times 6{,}530 = 5{,}126$ square mils

Example 3.3

Problem:

A rectangular conductor is 1.5 inches wide and 0.25 inch thick. (1) What is its area in square mils? (2) What size of round conductor in circular mils is necessary to carry the same current as the rectangular bar?

Solution:

1.

$$1.5'' = 1.5 \times 1,000 = 1,500 \text{ mils}$$

$$0.25'' = 0.25 \times 1,000 = 250 \text{ mils}$$

$$A = 1,500 \times 250 = 375,000 \text{ square mils}$$

2. To carry the same current, the cross-sectional area of the rectangular bar and the cross-sectional area of the round conductor must be equal. There are more circular mils than square mils in this area, and therefore

$$A = \frac{375,000}{0.785} = 477,700 \text{ circular mils}$$

Note: Many electric cables are composed of stranded wires. The strands are usually single wires twisted together in sufficient numbers to make up the necessary cross-sectional area of the cable. The total area in circular mils is determined by multiplying the area of one strand in circular mils by the number of strands in the cable.

Circular-Mil-Foot

As shown in Figure 3.5, a *circular-mil-foot* is actually a unit of volume. More specifically, it is a unit conductor 1 foot in length and having a cross-sectional area of 1 circular mil. The circular-mil-foot is useful in making comparisons between wires that are made of different metals because it is considered a unit conductor. Because it is considered a unit conductor, the circular-mil-foot is useful in making comparisons between wires that are made of different metals. For example, a basis of comparison of the **resistivity** of various substances may be made by determining the resistance of a circular-mil-foot of each of the substances.

Note: It is sometimes more convenient to employ a different unit of volume when working with certain substances. Accordingly, unit volume may also be taken as the centimeter cube. The inch cube may also be used. The unit of volume employed is given in tables of specific resistances.

Resistivity

All materials differ in their atomic structure and therefore in their ability to resist the flow of an electric current. The measure of the ability of a specific material to resist

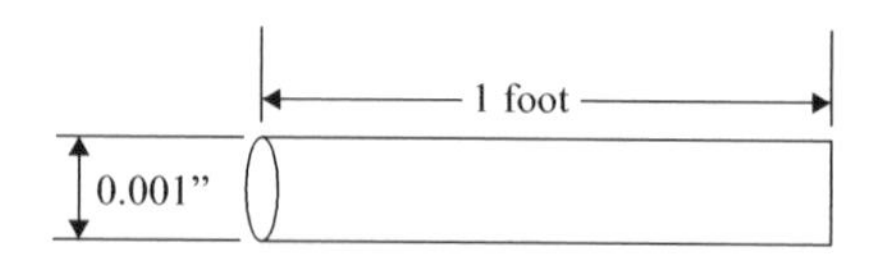

FIGURE 3.5 Circular-mil-foot.

the flow of electricity is called its *resistivity*, or specific resistance—the resistance in ohms offered by unit volume (the circular-mil-foot) of a substance to the flow of electric current. Resistivity is the reciprocal of conductivity (i.e., the ease by which current flows in a conductor). A substance that has a high resistivity will have a low conductivity, and vice versa.

The resistance of a given length, for any conductor, depends upon the resistivity of the material, the length of the wire, and the cross-sectional area of the wire according to the equation

$$R = \rho \frac{L}{A} \tag{3.3}$$

where

R = resistance of the conductor, Ω
L = length of the wire, ft
A = cross-sectional area of the wire, CM
ρ = specific resistance or resistivity, CM × Ω/ft

The factor ρ (Greek letter rho, pronounced "roe") permits different materials to be compared for resistance according to their nature without regard to different lengths or areas. Higher values of ρ mean more resistance.

Key Point: The resistivity of a substance is the resistance of a unit volume of that substance.

Many tables of resistivity are based on the resistance in ohms of a volume of the substance 1 foot long and 1 circular mil in cross-sectional area. The temperature at which the resistance measurement is made is also specified. If the kind of metal of which the conductor is made is known, the resistivity of the metal may be obtained from a table. The resistivity, or specific resistance, of some common substances is given in Table 3.1.

Note: Since silver, copper, gold, and aluminum have the lowest values of resistivity, they are the best conductors. Tungsten and iron have a much higher resistivity.

TABLE 3.1
Resistivity (Specific Resistance)

Substance	Specific Resistance @ 20° (CM ft Ω)
Silver	9.8
Copper (drawn)	10.37
Gold	14.7
Aluminum	17.02
Tungsten	33.2
Brass	42.1
Steel (soft)	95.8
Nichrome	660.0

Example 3.4

Problem:

What is the resistance of 1,000 feet of copper wire having a cross-sectional area of 10,400 circular mils (No. 10 wire), the wire temperature being 20°C?

Solution:

The resistivity (specific resistance), from Table 3.1, is 10.37. Substituting the known values in Equation (3.3), the resistance, R, is determined as

$$R = \rho \frac{L}{A} = 10.37 \times \frac{1{,}000}{10{,}400} = 1\ \Omega\text{, approximately}$$

Wire Measurement

Wires are manufactured in sizes numbered according to a table known as the American wire gage (AWG). Table 3.2 lists the standard wire sizes that correspond to the AWG. The gage numbers specify the size of round wire in terms of its diameter and cross-sectional area. Note the following:

a. As the gage numbers increase from 1 to 40, the diameter and circular area decrease. Higher gage numbers mean smaller wire sizes. Thus, No. 12 is a smaller wire than No. 4.
b. The circular area doubles for every three gage sizes. For example, No. 12 wire has about twice the area of No. 15 wire.
c. The higher the gage number and the smaller the wire, the greater the resistance of the wire for any given length. Therefore, for 1,000 feet of wire, No. 12 has a resistance of 1.62 Ω while No. 4 has 0.253 Ω.

Factors Governing Selection of Wire Size

Several factors must be considered in selecting the size of wire to be used for transmitting and distributing electric power. These factors include: allowable power loss in the line; the permissible voltage drop in the line; the current-carrying capacity of the line; and the ambient temperatures at which the wire is to be used.

a. **Allowable power loss (I^2R) in the line**—This loss represents electrical energy converted into heat. The use of large conductors will reduce the resistance and therefore the I^2R loss. However, large conductors are heavier and require more substantial supports; thus, they are more expensive initially than small ones.
b. **Permissible voltage drop (IR drop) in the line**—If the source maintains a constant voltage at the input to the line, any variation in the load on the line will cause a variation in line current, and a consequent variation in the IR drop in the line. A wide variation in the IR drop in the line causes poor voltage regulation at the load.

TABLE 3.2
Copper Wire Table

Gage #	Diameter	Circular mils	Ohms/1,000 ft @ 25°C
1	289.0	83,700.0	0.126
2	258.0	66,400.0	0.159
3	229.0	52,600.0	0.201
4	204.0	41,700.0	0.253
5	182.0	33,100.0	0.319
6	162.0	26,300.0	0.403
7	144.0	20,800.0	0.508
8	128.0	16,500.0	0.641
9	114.0	13,100.0	0.808
10	102.0	10,400.0	1.02
11	91.0	8,230.0	1.28
12	81.0	6,530.0	1.62
13	72.0	5,180.0	2.04
14	64.0	4,110.0	2.58
15	57.0	3,260.0	3.25
16	51.0	2,580.0	4.09
17	45.0	2,050.0	5.16
18	40.0	1,620.0	6.51
19	36.0	1,290.0	8.21
20	32.0	1,020.0	10.4
21	28.5	810.0	13.1
22	25.3	642.0	16.5
23	22.6	509.0	20.8
24	20.1	404.0	26.4
25	17.9	320.0	33.0
26	15.9	254.0	41.6
27	14.2	202.0	52.5
28	12.6	160.0	66.2
29	11.3	127.0	83.4
30	10.0	101.0	105.0
31	8.9	79.7	133.0
32	8.0	63.2	167.0
33	7.1	50.1	211.0
34	6.3	39.8	266.0
35	5.6	31.5	335.0
36	5.0	25.0	423.0
37	4.5	19.8	533.0
38	4.0	15.7	673.0
39	3.5	12.5	848.0
40	3.1	9.9	1,070.0

c. **The current-carrying capacity of the line**—When current is draw through the line, heat is generated. The temperature of the line will rise until the heat radiated, or otherwise dissipated, is equal to the heat generated by the passage of current through the line. If the conductor is insulated, the heat generated in the conductor is not so readily removed as it would be if the conductor were not insulated.
d. **Conductors installed where ambient temperature is relatively high**—When installed in such surroundings, the heat generated by external sources constitutes an appreciable part of the total conductor heating. Due allowance must be made for the influence of external heating on the allowable conductor current and each case has its own specific limitations.

Copper vs. Other Metal Conductors

If it were not cost prohibitive, silver, the best conductor of electron flow (electricity), would be the conductor of choice in electrical systems. Instead, silver is used only in special circuits where a substance with high conductivity is required.

The two most generally used conductors are copper and aluminum. Each has characteristics that make its use advantageous under certain circumstances. Likewise, each has certain disadvantages, or limitations.

In regard to **copper**, it has a higher conductivity; it is more ductile (can be drawn out into wire), has relatively high tensile strength, and can be easily soldered. It is more expensive and heavier than aluminum.

Aluminum has only about 60% of the conductivity of copper, but its lightness makes possible long spans, and its relatively large diameter for a given conductivity reduces corona (i.e., the discharge of electricity from the wire when it has a high potential). The discharge is greater when smaller diameter wire is used than when larger-diameter wire is used. However, aluminum conductors are not easily soldered, and aluminum's relatively large size for a given conductance does not permit the economical use of an insulation covering.

Note: Recent practice involves using copper wiring (instead of aluminum wiring) in house and some industrial applications. This is the case because aluminum connections are not as easily made as they are with copper. In addition, over the years, many fires have been started because of improperly connected aluminum wiring (i.e., poor connections = high-resistance connections, resulting in excessive heat generation).

A comparison of some of the characteristics of copper and aluminum is given in Table 3.3.

Temperature Coefficient (Wire)

The resistance of pure metals—such as silver, copper, and aluminum—increases as the temperature increases. The *temperature coefficient* of resistance, α (Greek letter alpha), indicates how much the resistance changes for a change in temperature. A positive value for α means R increases with temperature; a negative α means R

TABLE 3.3
Characteristics of Copper and Aluminum

Characteristics	Copper	Aluminum
Tensile strength (lb/in^2)	55,000	25,000
Tensile strength for same conductivity (lb)	55,000	40,000
Weight for same conductivity (lb)	100	48
Cross section for same conductivity (CM)	100	160
Specific resistance (Ω/mil. ft)	10.6	17

TABLE 3.4
Properties of Conducting Materials (Approximate)

Material	Temperature Coefficient, Ω/°C
Aluminum	0.004
Carbon	−0.0003
Constantan	0 (average)
Copper	0.004
Gold	0.004
Iron	0.006
Nichrome	0.0002
Nickel	0.005
Silver	0.004
Tungsten	0.005

decreases; and a zero α means R is constant, not varying with changes in temperature. Typical values of α are listed in Table 3.4.

The amount of increase in the resistance of a 1-Ω sample of the copper conductor per degree rise in temperature (i.e., the temperature coefficient of resistance) is approximately 0.004. For pure metals, the temperature coefficient of resistance ranges between 0.004 and 0.006 Ω.

Thus, a copper wire having a resistance of 50 Ω at an initial temperature of 0°C will have an increase in resistance of 50×0.004, or 0.2 Ω (approximate), for the entire length of wire for each degree of temperature rise above 0°C. At 20°C the increase in resistance is approximately 20×0.2, or 4 Ω. The total resistance at 20°C is $50 + 4$, or 54 Ω.

Note: As shown in Table 3.4, carbon has a negative temperature coefficient. In general, α is negative for all semiconductors such as germanium and silicon. A negative value for α means less resistance at higher temperatures. Therefore, the resistance of semiconductor diodes and transistors can be reduced considerably when they become hot with normal load current. Observe, also, that constantan has a value of zero for α (Table 3.4). Thus, it can be used for precision wire-wound resistors, which do not change resistance when the temperature increases.

Conductor Insulation

Electric current must be contained; it must be channeled from the power source to a useful load—safely. To accomplish this, electric current must be forced to flow only where it is needed. Moreover, current-carrying conductors must not be allowed (generally) to come in contact with one another, their supporting hardware, or personnel working near them. To accomplish this, conductors are coated or wrapped with various materials. These materials have such a high resistance that they are, for all practical purposes, nonconductors. They are generally referred to as *insulators* or *insulating materials*.

There are a wide variety of insulated conductors available to meet the requirements of any job. However, only the necessary minimum of insulation is applied for any particular type of cable designed to do a specific job. This is the case because insulation is expensive and has a stiffening effect and is required to meet a great variety of physical and electrical conditions.

Two fundamental but distinctly different properties of insulation materials (e.g., rubber, glass, asbestos, and plastics) are insulation resistance and dielectric strength.

a. *Insulation resistance* is the resistance to current leakage through and over the surface of insulation materials.
b. *Dielectric strength* is the ability of the insulator to withstand potential difference and is usually expressed in terms of the voltage at which the insulation fails because of the electrostatic stress.

Various types of materials are used to provide insulation for electric conductors, including rubber, plastics, varnished cloth, paper, silk, cotton, and enamel.

Conductor Splices and Terminal Connections

When conductors join each other, or connect to a load, *splices* or *terminals* must be used. It is important that they be properly made, since any electric circuit is only as good as its weakest connection. The basic requirement of any splice or connection is that it be both mechanically and electrically as strong as the conductor or device with which it is used. High-quality workmanship and materials must be employed to ensure lasting electrical contact, physical strength, and insulation (if required).

Important Point: Conductor splices and connections are an essential part of any electric circuit.

Soldering Operations

Soldering operations are a vital part of electrical and/or electronics maintenance procedures. Soldering is a manual skill that must be learned by all personnel who work in the field of electricity. Obviously, practice is required to develop proficiency in the techniques of soldering.

In performing a soldering operation both the solder and the material to be soldered (e.g., electric wire and/or terminal lugs) must be heated to a temperature that allows the solder to flow. If either is heated inadequately, **cold** solder joints result (i.e., high-resistance connections are created). Such joints do not provide either the physical strength or the electrical conductivity required. Moreover, in soldering operations, it is necessary to select a solder that will flow at a temperature low enough to avoid damage to the part being soldered, or to any other part or material in the immediate vicinity.

Solderless Connectors

Generally, terminal lugs and splicers that do not require solder are more widely used (because they are easier to mount correctly) than those that do require solder. Solderless connectors—made in a wide variety of sizes and shapes—are attached to their conductors by means of several different devices, but the principle of each is essentially the same. They are all crimped (squeezed) tightly onto their conductors. They afford adequate electrical contact, plus great mechanical strength.

Insulation Tape

The carpenter has his saw, the dentist his pliers, the plumber his wrench, and the electrician his insulating tape. Accordingly, one of the first things the rookie maintenance operator learns (a rookie who is also learning proper and safe techniques for performing electrical work) is the value of electrical insulation tape. Normally, the use of electrical insulating tape comes into play as the final step in completing a splice or joint, to place insulation over the bare wire at the connection point.

Typically, insulation tape used should be the same basic substance as the original insulation, usually a rubber-splicing compound. When using rubber (latex) tape as the splicing compound where the original insulation was rubber, it should be applied to the splice with a light tension so that each layer presses tightly against the one underneath it. In addition to the rubber tape application (which restores the insulation to original form), restoring with friction tape is also often necessary.

In recent years, plastic electrical tape has come into wide use. It has certain advantages over rubber and friction tape. For example, it will withstand higher voltages for a given thickness. Single thin layers of certain commercially available plastic tape will stand several thousand volts without breaking down.

Important Point: Be advised that though the use of plastic electrical tape has become almost universal in industrial applications, it must be applied in more layers—because it is thinner than rubber or friction tape—to ensure an extra margin of safety.

Static Electricity

Electricity at rest is often referred to as *static electricity*. More specifically, when two bodies of matter have unequal charges, and are near one another, an electric force is exerted between them because of their unequal charges. However, since they are not in contact, their charges cannot equalize. The existence of such an electric force, where current can't flow, is *static electricity*.

However, static electricity, or electricity at rest, will flow if given the opportunity—this is the case because static electricity is an imbalance of negative and positive charges. An example of this phenomenon is often experienced when one walks across a dry carpet and then touches a doorknob—a slight shock is usually felt and a spark at the fingertips is likely noticed. Another familiar example is "static cling." For example, whenever we rub an air-filled balloon against the hair on our heads, and then place the balloon against a wall the balloon sticks to the wall, defying gravity, because of static cling. In the workplace, static electricity is prevented from building up by properly bonding or grounding of equipment to ground or earth.

Charged Bodies

The fundamental law of charged bodies states: "Like charges repel each other and unlike charges attract each other." A positive charge and negative charge, being opposite or unlike, tend to move toward each other—attracting each other. In contrast, like bodies tend to repel each other. Electrons repel each other because of their like negative charges, and protons repel each other because of their like positive charges. Figure 3.6 demonstrates the law of charged bodies.

It is important to point out another significant part of the fundamental law of charged bodies—that is, *force of attraction or repulsion existing between two magnetic poles decreases rapidly as the poles are separated from each other*. More specifically, the force of attraction or repulsion varies directly as the product of the separate pole strengths and inversely as the square of the distance separating the magnetic poles, provided the poles are small enough to be considered as points.

Let's look at an example:

If you increase the distance between two north poles of magnets from 2 to 4 feet, the force of repulsion between them is decreased to one-fourth of its original value. If either pole strength is doubled, the distance remaining the same, the force between the poles will be doubled.

Coulomb's Law

Simply put, Coulomb's Law states that the amount of attracting or repelling force that acts between two electrically charged bodies in free space depends on two things:

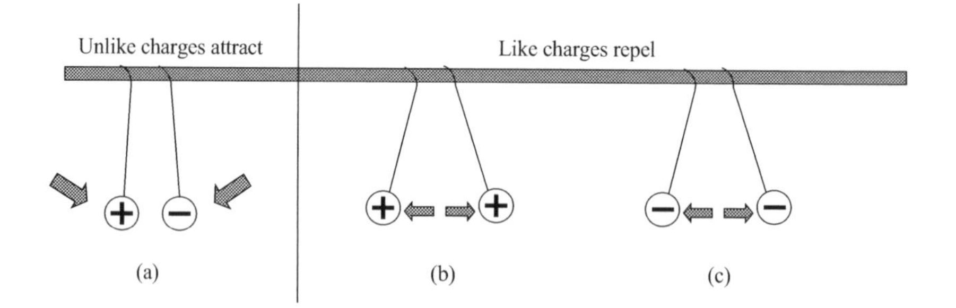

FIGURE 3.6 Reaction between two charged bodies. The opposite charge in (a) attracts. The like charges in (b) and (c) repel each other.

a. Their charges
b. The distance between them.

Specifically, *Coulomb's Law* states: "Charged bodies attract or repel each other with a force that is directly proportional to the product of their charges and is inversely proportional to the square of the distance between them."

Note: The magnitude of electric charge a body possesses is determined by the number of electrons compared with the number of protons within the body. The symbol for the magnitude of electric charge is Q, expressed in units of *coulombs* (C). A charge of one positive coulomb means a body contains a charge of 6.25×10^{18}. A charge of one negative coulomb, $-Q$, means a body contains a charge of 6.25×10^{18} more electrons than protons.

Electrostatic Fields

The fundamental characteristic of an electric charge is its ability to exert force. The space between and around charged bodies in which their influence is felt is called an *electric field of force*. The electric field is always terminated on material objects and extends between positive and negative charges. This region of force can consist of air, glass, paper, or a vacuum. This region of force is referred to as an *electrostatic field*.

When two objects of opposite polarity are brought near each other, the electrostatic field is concentrated in the area between them. The field is generally represented by lines that are referred to as *electrostatic lines of force*. These lines are imaginary and are used merely to represent the direction and strength of the field. To avoid confusion, the positive lines of force are always shown leaving charge, and for a negative charge they are shown as entering. Figure 3.7 illustrates the use of lines to represent the field about charged bodies.

Note: A charged object will retain its charge temporarily if there is no immediate transfer of electrons to or from it. In this condition, the charge is said to be *at rest*. Remember, electricity at rest is called *static* electricity.

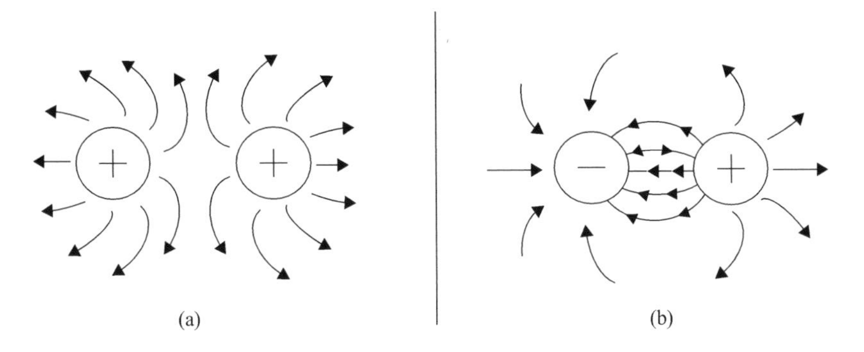

FIGURE 3.7 Electrostatic lines of force. (a) represents the repulsion of like-charged bodies and their associated fields. (b) represents the attraction between unlike-charged bodies and their associated fields.

Magnetism

Most electrical equipment depends directly or indirectly upon magnetism. Magnetism is defined as a phenomenon associated with magnetic fields; that is, it has the power to attract such substances as iron, steel, nickel, or cobalt (metals that are known as magnetic materials). Correspondingly, a substance is said to be a magnet if it has the property of magnetism. For example, a piece of iron can be magnetized and is thus a magnet.

When magnetized, the piece of iron (note: we will assume a piece of flat bar 6 inches long × 1 inch wide × 0.5 inch thick; a bar magnet—see Figure 3.8) will have two points opposite each other, which most readily attract other pieces of iron. The points of maximum attraction (one on each end) are called the *magnetic poles* of the magnet: the north (N) pole and the south (S) pole. Just as like electric charges repel each other and opposite charges attract each other, like magnetic poles repel each other and unlike poles attract each other. Although invisible to the naked eye, its force can be shown to exist by sprinkling small iron filings on a glass covering a bar magnet as shown in Figure 3.8.

Figure 3.9 shows how the field looks without iron filings; it is shown as lines of force [known as *magnetic flux or flux lines*; the symbol for magnetic flux is the Greek lowercase letter ϕ (phi)] in the field, repelled away from the north pole of the magnet and attracted to its south pole.

Note: A *magnetic circuit* is a complete path through which magnetic lines of force may be established under the influence of a magnetizing force. Most magnetic circuits are composed largely of magnetic materials in order to contain the magnetic flux. These circuits are similar to the *electric circuit* (an important point), which is

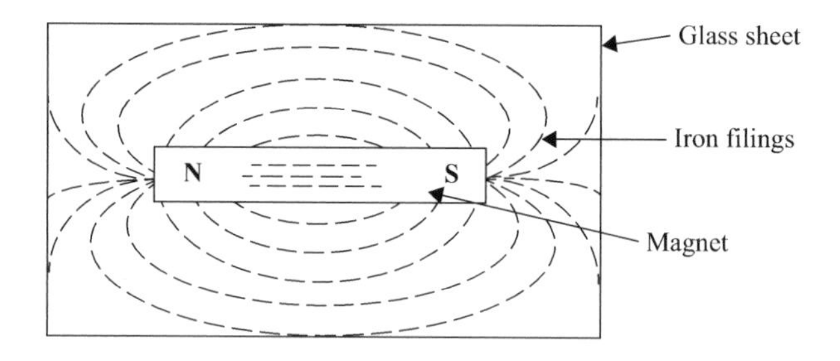

FIGURE 3.8 The magnetic field around a bar magnet. If the glass sheet is tapped gently, the filings will move into a definite pattern that describes the field of force around the magnet.

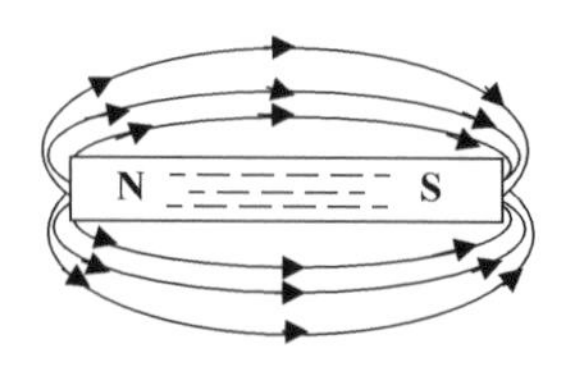

FIGURE 3.9 Magnetic field of force around a bar magnet, indicated by lines of force.

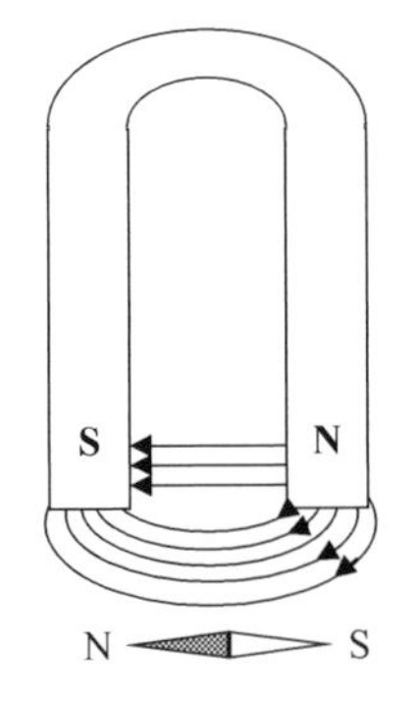

FIGURE 3.10 Horseshoe magnet.

a complete path through which current is caused to flow under the influence of an electromotive force.

There are three types or groups of magnets:

a. **Natural magnets**—found in the natural state in the form of a mineral (an iron compound) called magnetite.
b. **Permanent magnets**—(artificial magnet) hardened steel or some alloy such as Alnico bars that have been permanently magnetized. The permanent magnet most people are familiar with is the horseshoe magnet; this red U-shaped magnet is the universal symbol of magnets, recognized throughout the world (see Figure 3.10).
c. **Electromagnets**—(artificial magnet) composed of soft iron cores around which are wound coils of insulated wire. When an electric current flows through the coil, the core becomes magnetized. When the current ceases to flow, the core loses most of the magnetism.

Magnetic Materials

Natural magnets are no longer used (they have no practical value) in electrical circuitry because more powerful and more conveniently shaped permanent magnets can be produced artificially. Commercial magnets are made from special steels and alloys—magnetic materials.

Magnetic materials are those materials that are attracted or repelled by a magnet and that can be magnetized themselves. Iron, steel, and alloy bar are the most common magnetic materials. These materials can be magnetized by inserting the material (in bar form) into a coil of insulated wire and passing a heavy direct current through the coil. The same material may also be magnetized if it is stroked with a bar magnet. It will then have the same magnetic property that the magnet used to induce the magnetism has—namely, there will be two poles of attraction, one at either end. This process produces a permanent magnet by induction—that is, the magnetism is induced in the bar by the influence of the stroking magnet.

Note: Permanent magnets are those of hard magnetic materials (hard steel or alloys) that retain their magnetism when the magnetizing field is removed.

A temporary magnet is one that has *no* ability to retain a magnetized state when the magnetizing field is removed.

Even though classified as permanent magnets, it is important to point out that hardened steel and certain alloys are relatively difficult to magnetize and are said to have a *low permeability* because the magnetic lines of force do not easily permeate or distribute themselves readily through the steel.

Note: *Permeability* refers to the ability of a magnetic material to concentrate magnetic flux. Any material that is easily magnetized has high permeability. A measure of permeability for different materials in comparison with air or vacuum is called *relative* permeability, symbolized by μ or (mu).

Once hard steel and other alloys are magnetized, however, they retain a large part of their magnetic strength and are called *permanent magnets*. Conversely, materials that are relatively easy to magnetize—such as soft iron and annealed silicon steel—are said to have a *high permeability*. Such materials retain only a small part of their magnetism after the magnetizing force is removed and are called *temporary magnets*.

The magnetism that remains in a temporary magnet after the magnetizing force is removed is called *residual magnetism*.

Early magnetic studies classified magnetic materials merely as being magnetic and nonmagnetic—that is, based on the strong magnetic properties of iron. However, since weak magnetic materials can be important in some applications, present studies classify materials into one of three groups: namely, paramagnetic, diamagnetic, and ferromagnetic.

a. **Paramagnetic materials**—These include aluminum, platinum, manganese, and chromium—materials that become only slightly magnetized even though under the influence of a strong magnetic field. This slight magnetization is in the same direction as the magnetizing field. Relative permeability is slightly more than 1 (i.e., considered nonmagnetic materials).
b. **Diamagnetic materials**—These include bismuth, antimony, copper, zinc, mercury, gold, and silver—materials that can also be slightly magnetized when under the influence of a very strong field. Relative permeability is less than 1 (i.e., considered nonmagnetic materials).
c. **Ferromagnetic materials**—These include iron, steel, nickel, cobalt, and commercial alloys—materials that are the most important group for applications of electricity and electronics. Ferromagnetic materials are easy to magnetize and have high permeability, ranging from 50 to 3,000.

Magnetic Earth

The earth is a huge magnet; and surrounding the earth is the magnetic field produced by the earth's magnetism. Most people would have no problem understanding or at least accepting this statement. However, if told that the earth's north magnetic pole is actually its south magnetic pole and that the south magnetic pole is actually the earth's north magnetic pole, they might not accept or understand this statement. But, in terms of a magnet, it is true.

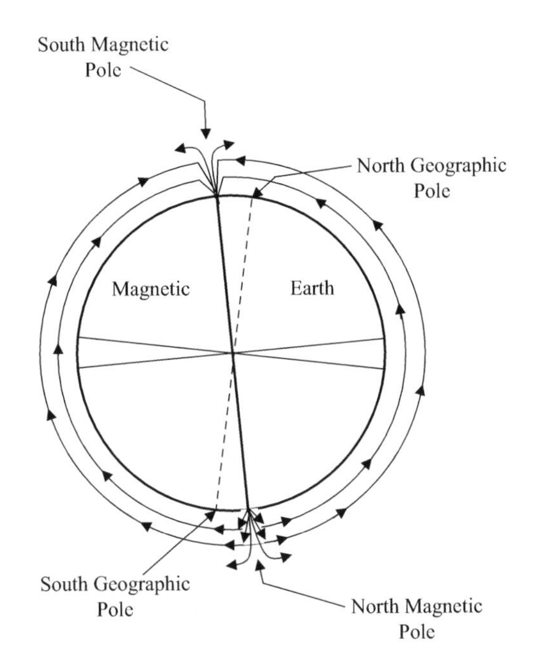

FIGURE 3.11 Earth's magnetic poles.

As shown in Figure 3.11, the geographic poles are also shown at each end of the axis of rotation of the earth. Clearly, as shown in Figure 3.11, the magnetic axis does not coincide with the geographic axis, and therefore the magnetic and geographic poles are not at the same place on the surface of the earth.

Recall that magnetic lines of force are assumed to emanate from the north pole of a magnet and to enter the south pole as closed loops. Because the earth is a magnet, lines of force emanate from its north magnetic pole and enter the south magnetic pole as closed loops. A compass needle aligns itself in such a way that the earth's lines of force enter at its south pole and leave at its north pole. Because the north pole of the needle is defined as the end that points in a northerly direction it follows that the magnetic pole in the vicinity of the north geographic pole is in reality a south magnetic pole, and vice versa.

Difference in Potential

Because of the force of its electrostatic field, an electric charge has the ability to do the work of moving another charge by attraction or repulsion. The force that causes free electrons to move in a conductor as an electric current may be referred to as follows:

- Electromotive force (emf)
- Voltage
- Difference in potential

When a difference in potential exists between two charged bodies that are connected by a wire (conductor), electrons (current) will flow along the conductor. This flow is from the negatively charged body to the positively charged body until the two charges are equalized and the potential difference no longer exists.

Note: The basic unit of potential difference is the *volt* (V). The symbol for potential difference is *V*, indicating the ability to do the work of forcing electrons (current flow) to move. Because the volt unit is used, potential difference is called *voltage*.

Water Analogy

In attempting to train individuals in the concepts of basic electricity, especially in regard to difference of potential (voltage), current, and resistance relationships in a simple electrical circuit, it has been common practice to use what is referred to as the water analogy. We use the water analogy later to explain (in a simple straightforward fashion) voltage, current, and resistance and their relationships in more detail, but for now we use the analogy to explain the basic concept of electricity: difference of potential, or voltage. Because a difference in potential causes current flow (against resistance), it is important that this concept be understood first before the concepts of current flow and resistance are explained.

Consider the water tanks shown in Figure 3.12—two water tanks connected by a pipe and valve. At first, the valve is closed and all the water is in Tank A. Thus, the water pressure across the valve is at maximum. When the valve is opened, the water flows through the pipe from A to B until the water level becomes the same in both tanks. The water then stops flowing in the pipe, because there is no longer a difference in water pressure (difference in potential) between the two tanks.

Just as the flow of water through the pipe in Figure 3.12 is directly proportional to the difference in water level in the two tanks, current flow through an electric circuit is directly proportional to the difference in potential across the circuit.

Important Point: A fundamental law of current electricity is that the current is directly proportional to the applied voltage; that is, if the voltage is increased, the current is increased. If the voltage is decreased, the current is decreased.

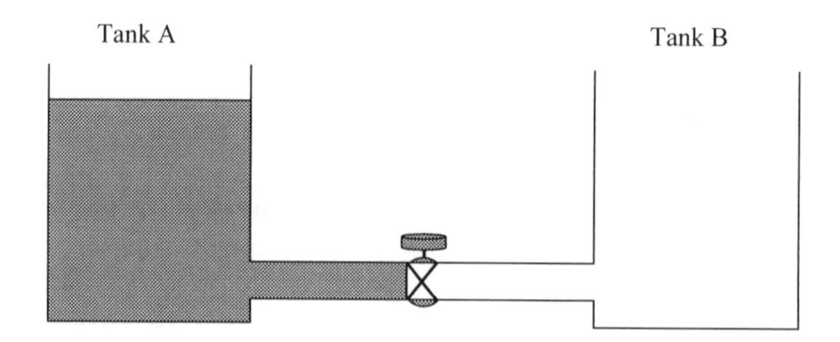

FIGURE 3.12 Water analogy of electric difference of potential.

Principal Methods of Producing a Voltage

There are many ways to produce electromotive force, or voltage. Some of these methods are much more widely used than others. The following is a list of the seven most common methods of producing electromotive force (USDOE 1992).

a. **Friction**—voltage produced by rubbing two materials together (static electricity or electrostatic force). Remember our discussion of static electricity? Let's refresh our memories. For example, have you ever walked across a carpet and received a shock when you touched a metal doorknob? Your shoe soles built up a charge by rubbing on the carpet, and this charge was transferred to your body. Your body became positively charged and, when you touched the zero-charged doorknob, electrons were transferred to your body until both you and the doorknob had equal charges.
b. **Pressure (Piezoelectricity)**—voltage produced by squeezing or applying pressure to crystals of certain substances (e.g., crystals like quartz or Rochelle salts or certain ceramics like barium titanate). When pressure is applied to such substances, electrons can be driven out of orbit in the direction of the force. Electrons leave one side of the material and accumulate on the other side, building up positive and negative charges on opposite sides. When the pressure is released, the electrons return to their orbits. Some materials will react to bending pressure, while others will respond to twisting pressure. This generation of voltage is known as the *piezoelectric effect*. If external wires are connected while pressure and voltage are present, electrons will flow and current will be produced. If the pressure is held constant, the current will flow until the potential difference is equalized. When the force is removed, the material is decompressed and immediately causes an electric force in the opposite direction. The power capacity of these materials is extremely small. However, these materials are very useful because of their extreme sensitivity to changes of mechanical force. One example is the crystal phonograph cartridge that contains a Rochelle salt crystal. A phonograph needle is attached to the crystal. As the needle moves in the grooves of a record, it swings from side to side, applying compression and decompression to the crystal. This mechanical motion applied to the crystal generates a voltage signal that is used to reproduce sound.
c. **Heat (Thermoelectricity)**—voltage produced by heating the joint (junction) where two unlike metals are joined. Some materials readily give up their electrons and others readily accept electrons. For example, when two dissimilar metals like copper and zinc are joined together, a transfer of electrons can take place. Electrons will leave the copper atoms and enter the zinc atoms. The zinc gets a surplus of electrons and becomes negatively charged. The copper loses electrons and takes on a positive charge. This creates a voltage potential across the junction of the two metals. The heat energy of normal room temperature is enough to make them release and gain electrons, causing a measurable voltage potential. As more heat

energy is applied to the junction, more electrons are released, and the voltage potential becomes greater. When heat is removed, the junction cools, the charges will dissipate, and the voltage potential will decrease. This process is called thermoelectricity. A device like this is generally referred to as a thermocouple.

The voltage in a thermocouple is dependent upon the heat energy applied to the junction of the two dissimilar metals. Thermocouples are widely used to measure temperature and as heat-sensing devices in automatic temperature-controlled equipment.

Thermocouple power capacities are very small compared to some other sources but are somewhat greater than those of crystals. Generally speaking, a thermocouple can be subjected to higher temperatures than ordinary mercury or alcohol thermometers.

d. **Light (Photoelectricity)**—voltage produced by light (photons) striking photosensitive (light-sensitive) substances. When the photons in a light beam strike the surface of a material, they release their energy and transfer it to the atomic electrons of the material. This energy transfer may dislodge electrons from their orbits around the surface of the substance. Upon losing electrons, the photosensitive (light-sensitive) material becomes positively charged and an electric force is created.

This phenomenon is called the photoelectric effect and has wide applications in electronics, such as photoelectric cells, photovoltaic cells, optical couplers, and television camera tubes. Three uses of the photoelectric effect are described below.

- Photovoltaic: The light energy in one of two plates that are joined together causes one plate to release electrons to the other. The plates build up opposite charges, like a battery.
- Photoemission: The photon energy from a beam of light could cause a surface to release electrons in a vacuum tube. A plate would then collect the electrons.
- Photoconduction: The light energy applied to some materials that are normally poor conductors causes free electrons to be produced in the materials so that they become better conductors.

e. **Chemical action**—voltage produced by chemical reaction in a battery cell. For example, the voltaic chemical cell wherein a chemical reaction produces and maintains opposite charges on two dissimilar metals that serve as positive and negative terminals. The metals are in contact with an electrolyte solution. Connecting together more than one of these cells will produce a battery.

f. **Magnetism**—voltage produced in a conductor when the conductor moves through a magnetic field, or a magnetic field moves through the conductor in such a manner as to cut the magnetic lines of force of the field. A generator is a machine that converts mechanical energy into electrical energy by using the principle of *magnetic induction*. This is one of the most useful and widely employed applications of producing vast quantities of electric power.

g. **Thermionic emission**—A thermionic energy converter is a device consisting of two electrodes placed near one another in a vacuum. One electrode is normally called the cathode, or emitter, and the other is called the anode, or plate. Ordinarily electrons in the cathode are prevented from escaping from the surface by a potential energy barrier. When an electron starts to move away from the surface, it induces a corresponding positive charge in the material, which tends to pull it back into the surface. To escape, the electron must somehow acquire enough energy to overcome this energy barrier. At ordinary temperatures, almost none of the electrons can acquire enough energy to escape. However, when the cathode is very hot, the electron energies are greatly increased by thermal motion. At sufficiently high temperature, a considerable number of electrons are able to escape. The liberation of electrons from a hot surface is called *thermionic emission*.

 The electrons that have escaped from the hot cathode form a cloud of negative charges near it called a space charge. If the plate is maintained positive with respect to the cathode by a battery, the electrons in the cloud are attracted to it. As long as the potential difference between the electrodes is maintained, there will be a steady current flow from the cathode to the plate.

 The simplest example of a thermionic device is a vacuum tube diode in which the only electrodes are the cathode and plate, or anode. The diode can be used to convert alternating current (AC) flow to a pulsating direct current (DC) flow.

In the study of the basic electricity related to renewable energy production, we are most concerned with magnetism (generators powered by hydropower, for example), light (photoelectricity produced by solar cells), and chemistry (chemical energy converted to electricity in batteries) as means to produce voltage. Friction has little practical applications, though we discussed it earlier in studying static electricity. Pressure and heat do have useful applications, but we do not need to consider them in this text. Magnetism used in generators, solar light-produced electricity, and the chemistry involved with storing electricity in batteries, on the other hand, are, as mentioned, the principal sources of voltage and are discussed at length in this text.

Electric Current

The movement or the flow of electrons is called *current*. To produce current, the electrons must be moved by a potential difference or pressure (voltage).

Note: The terms current, current flow, electron flow, electron current, etc., may be used to describe the same phenomenon.

For our purposes in this text, electron flow, or current, in an electric circuit is from a region of less negative potential to a region of more positive potential—from negative to positive.

Note: Current is represented by the letter *I*. The basic unit in which current is measured is the *ampere, or amp (A)*. One ampere of current is defined as the movement of 1 coulomb past any point of a conductor during 1 second of time.

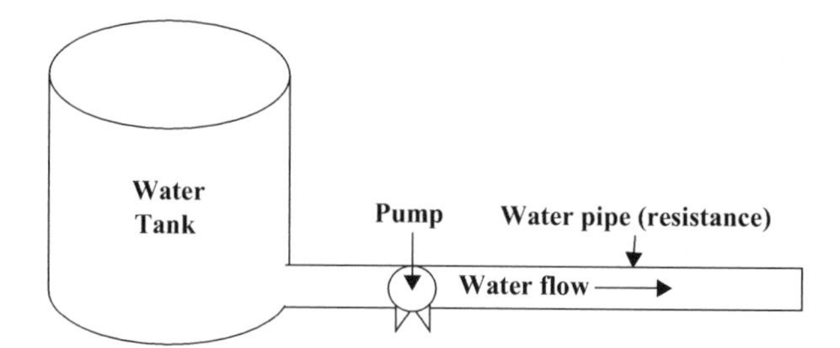

FIGURE 3.13 Water analogy: current flow.

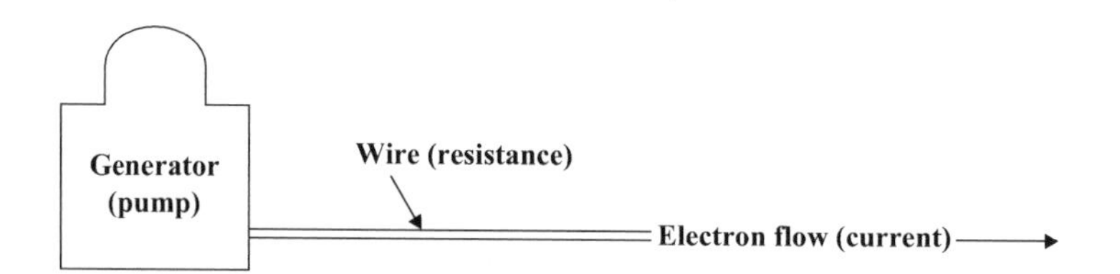

FIGURE 3.14 Simple electric circuit with current flow.

Recall that we used the water analogy to help us understand potential difference. We can also use the water analogy to help us understand current flow through a simple electric circuit. Consider Figure 3.13 that shows a water tank connected via a pipe to a pump with a discharge pipe. If the water tank contains an amount of water above the level of the pipe opening to the pump, the water exerts pressure (a difference in potential) against the pump. When sufficient water is available for pumping with the pump, water flows through the pipe against the resistance of the pump and pipe. The analogy should be clear—in an electric circuit if a difference of potential exists, current will flow in the circuit.

Another simple way of looking at this analogy is to consider Figure 3.13 where the water tank has been replaced with a generator, the pipe with a conductor (wire), and water flow with the flow of electric current.

Again, the key point illustrated by Figures 3.13 and 3.14 is that to produce current, the electrons must be moved by a potential difference.

Electric current is generally classified into two general types:

- direct current (d-c)
- alternating current (a-c)

Direct current is current that moves through a conductor or circuit in one direction only. *Alternating current* periodically reverses direction.

Resistance

Earlier it was pointed out that free electrons, or electric current, could move easily through a good conductor, such as copper, but that an insulator, such as glass, was an obstacle to current flow. In the water analogy shown in Figure 3.13 and the simple electric circuit shown in Figure 3.14, resistance is indicated by either the pipe or the conductor.

Every material offers some resistance, or opposition, to the flow of electric current through it. Good conductors such as copper, silver, and aluminum offer very little resistance. Poor conductors, or insulators, such as glass, wood, and paper, offer a high resistance to current flow.

Note: The amount of current that flows in a given circuit depends on two factors: voltage and resistance.

Note: Resistance is represented by the letter R. The basic unit in which resistance is measured is the *ohm* (Ω). One ohm is the resistance of a circuit element, or circuit, that permits a steady current of 1 ampere (1 coulomb per second) to flow when a steady electromotive force (emf) of 1 volt is applied to the circuit. Manufactured circuit parts containing definite amounts of resistance are called **resistors**.

The size and type of material of the wires in an electric circuit are chosen so as to keep the electrical resistance as low as possible. In this way, current can flow easily through the conductors, just as water flows through the pipe between the tanks in Figure 3.13. If the water pressure remains constant the flow of water in the pipe will depend on how far the valve is opened. The smaller the opening, the greater the opposition (resistance) to the flow, and the smaller will be the rate of flow in gallons per second.

In the electric circuit shown in Figure 3.14, the larger the diameter of the wire, the lower will be its electrical resistance (opposition) to the flow of current through it. In the water analogy, pipe friction opposes the flow of water between the tanks. This friction is similar to electrical resistance. The resistance of the pipe to the flow of water through it depends upon (1) the length of the pipe, (2) diameter of the pipe, and (3) the nature of the inside walls (rough or smooth). Similarly, the electrical resistance of the conductors depends upon (1) the length of the wires, (2) the diameter of the wires, and (3) the material of the wires (copper, silver, etc.).

It is important to note that temperature also affects the resistance of electrical conductors to some extent. In most conductors (copper, aluminum, etc.), the resistance increases with temperature. Carbon is an exception. In carbon the resistance decreases as temperature increases.

Important Note: Electricity is a study that is frequently explained in terms of opposites. The term that is exactly the opposite of resistance is *conductance*. Conductance (G) is the ability of a material to pass electrons. The SI derived unit of conductance is the siemens. The commonly used unit of conductance is the *Mho*, which is ohm spelled backward. The relationship that exists between resistance and conductance is the reciprocal. A reciprocal of a number is obtained by dividing the number by one. If the resistance of a material is known, dividing its value by one will give its conductance. Similarly, if the conductance is known, dividing its value by one will give its resistance.

ELECTROMAGNETISM

Earlier, fundamental theories concerning simple magnets and magnetism were presented. Those discussions dealt mainly with forms of magnetism that were not related directly to electricity—permanent magnets for instance. Further, only brief mention was made of those forms of magnetism having direct relation to electricity—producing electricity with magnetism for instance.

In medicine, anatomy and physiology are so closely related that the medical student cannot study one at length without involving the other. A similar relationship holds for the electrical field; that is, magnetism and basic electricity are so closely related that one cannot be studied at length without involving the other. This close fundamental relationship is continually borne out in the study of generators, transformers, battery packs, and motors. To be proficient in electricity, we must become familiar with such general relationships that exist between magnetism and electricity as follows:

a. Electric current flow will always produce some form of magnetism.
b. Magnetism is by far the most commonly used means for producing or using electricity.
c. The peculiar behavior of electricity under certain conditions is caused by magnetic influences.

Magnetic Field Around a Single Conductor

In 1819, Hans Christian Oersted, a Danish scientist, discovered that a field of magnetic force exists around a single wire conductor carrying an electric current. In Figure 3.15, a wire is passed through a piece of cardboard and connected through a switch to a dry cell. With the switch open (no current flowing), if we sprinkle iron filings on the cardboard then tap it gently, the filings will fall back haphazardly. Now, if we close the switch, current will begin to flow in the wire. If we tap the cardboard again, the magnetic effect of the current in the wire will cause the filings to fall back into a definite pattern of concentric circles with the wire as the center of the circles. Every section of the wire has this field of force around it in a plane perpendicular to the wire, as shown in Figure 3.16.

The ability of the magnetic field to attract bits of iron depends on the number of lines of force present. The strength of the magnetic field around a wire carrying a

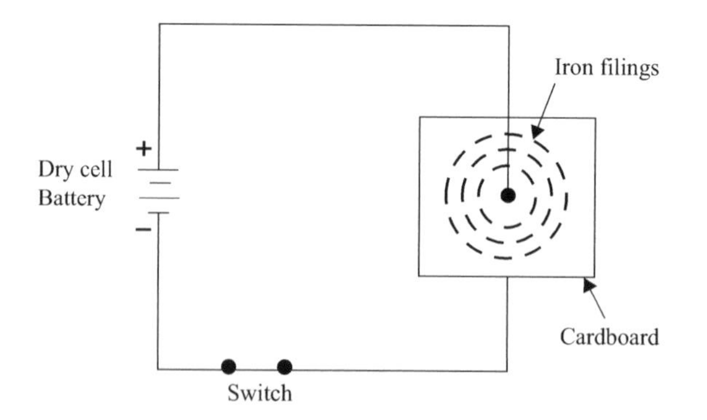

FIGURE 3.15 A circular pattern of magnetic force exists around a wire carrying an electric current.

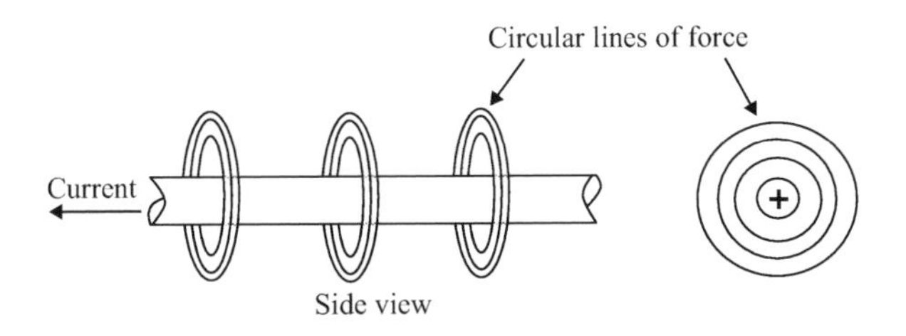

FIGURE 3.16 The circular fields of force around a wire carrying a current are in planes which are perpendicular to the wire.

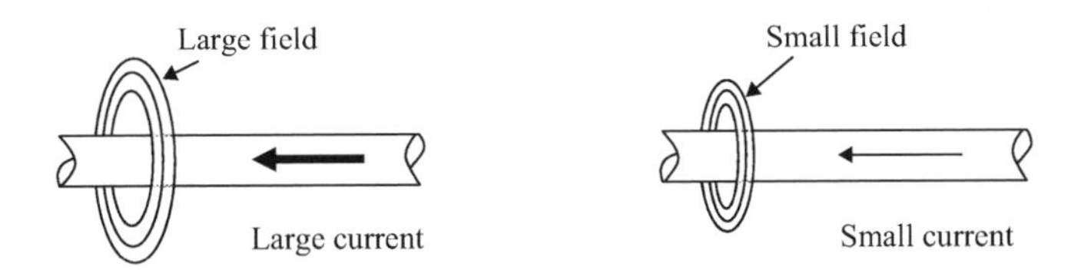

FIGURE 3.17 The strength of the magnetic field around a wire carrying a current depends on the amount of current.

current depends on the current, since it is the current that produces the field. The greater the current, the greater the strength of the field. A large current will produce many lines of force extending far from the wire, while a small current will produce only a few lines close to the wire, as shown in Figure 3.17.

Polarity of a Single Conductor

The relation between the direction of the magnetic lines of force around a conductor and the direction of current flow along the conductor may be determined by means of the **left-hand rule for a conductor**. If the conductor is grasped in the left hand with the thumb extended in the direction of electron flow (– to +), the fingers will point in the direction of the magnetic lines of force. This is the same direction that the north pole of a compass would point if the compass were placed in the magnetic field.

Important Note: Arrows are generally used in electric diagrams to denote the direction of current flow along the length of wire. Where cross sections of wire are shown, a special view of the arrow is used. A cross-sectional view of a conductor that is carrying current toward the observer is illustrated in Figure 3.18a. The direction of current is indicated by a dot, which represents the head of the arrow. A conductor that is carrying current away from the observer is illustrated in Figure 3.18b. The direction of current is indicated by a cross, which represents the tail of the arrow.

Field Around Two Parallel Conductors

When two parallel conductors carry current in the same direction, the magnetic fields tend to encircle both conductors, drawing them together with a force of attraction, as shown in Figure 3.19a. Two parallel conductors carrying currents in opposite

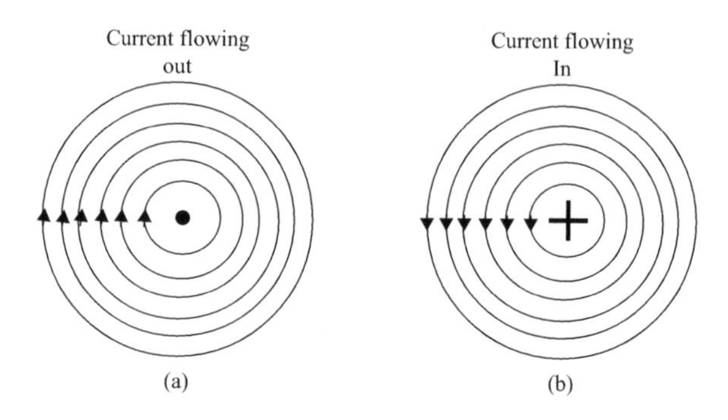

FIGURE 3.18 (a and b) Magnetic field around a current-carrying conductor.

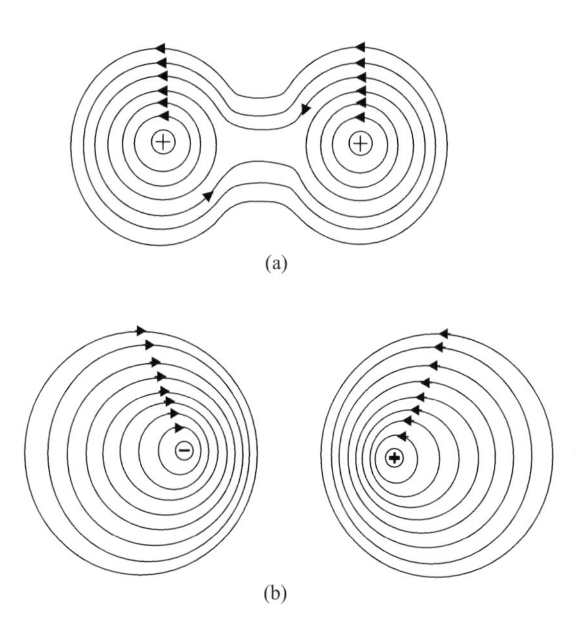

FIGURE 3.19 Magnetic field around two parallel conductors. (a) Current flowing in the same direction. (b) Current flowing in opposite directions.

directions are shown in Figure 3.19b. The field around one conductor is opposite in direction to the field around the other conductor. The resulting lines of force are crowded together in the space between the wires and tend to push the wires apart. Therefore, two parallel adjacent conductors carrying currents in the same direction attract each other and two parallel conductors carrying currents in opposite directions repel each other.

Magnetic Field of a Coil

The magnetic field around a current-carrying wire exists at all points along its length. Bending the current-carrying wire into the form of a single loop has two results.

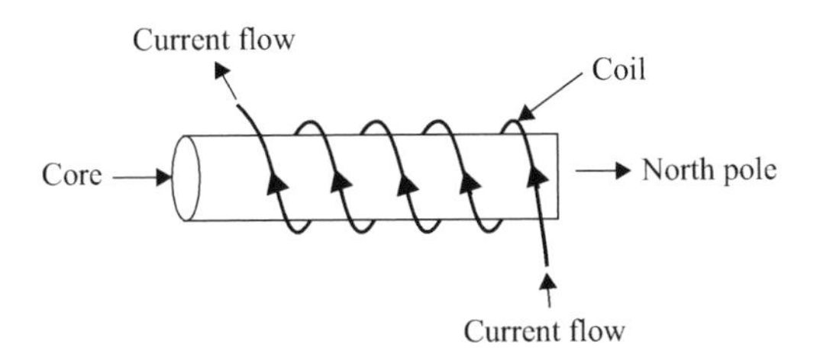

FIGURE 3.20 Current-carrying coil.

First, the magnetic field consists of more dense concentric circles in a plane perpendicular to the wire (see Figure 3.18), although the total number of lines is the same as for the straight conductor. Second, all the lines inside the loop are in the same direction. When this straight wire is wound around a core, as shown in Figure 3.20, it becomes a coil and the magnetic field assumes a different shape. When current is passed through the coiled conductor, the magnetic field of each turn of wire links with the fields of adjacent turns. The combined influence of all the turns produces a two-pole field similar to that of a simple bar magnet. One end of the coil will be a north pole and the other end will be a south pole.

Polarity of an Electromagnetic Coil

In Figure 3.18, it was shown that the direction of the magnetic field around a straight conductor depends on the direction of current flow through that conductor. Thus, a reversal of current flow through a conductor causes a reversal in the direction of the magnetic field that is produced. It follows that a reversal of the current flow through a coil also causes a reversal of its two-pole field. This is true because that field is the product of the linkage between the individual turns of wire on the coil. Therefore, if the field of each turn is reversed, it follows that the total field (coils' field) is also reversed.

When the direction of electron flow through a coil is known, its polarity may be determined by use of the **left-hand rule for coils**. This rule is illustrated in Figure 3.20 and is stated as follows: Grasping the coil in the left hand, with the fingers "wrapped around" in the direction of electron flow, the thumb will point toward the north pole.

Strength of an Electromagnetic Field

The strength, or intensity, of the magnetic field of a coil depends on a number of factors:

- The *number of turns* of conductor.
- The *amount of current flow* through the coil.
- The *ratio of the coil's length to its width.*
- The *type of material in the core.*

Magnetic Units

The law of current flow in the electric circuit is similar to the law for the establishing of flux in the magnetic circuit.

The *magnetic flux*, ϕ (phi), is similar to current in the Ohm's Law formula and comprises the total number of lines of force existing in the magnetic circuit. The **Maxwell** is the unit of flux—that is, 1 line of force is equal to 1 maxwell.

Note: The maxwell is often referred to as simply a line of force, line of induction, or line.

The *strength* of a magnetic field in a coil of wire depends on how much current flows in the turns of the coil. The more current, the stronger the magnetic field. Also, the more turns, the more concentrated are the lines of force. The *force* that produces the flux in the magnetic circuit (comparable to electromotive force in Ohm's Law) is known as *magnetomotive force*, or mmf. The practical unit of magnetomotive force is the **ampere-turn** (At). In equation form,

$$F = \text{ampere-turns} = NI \tag{3.4}$$

where

F = magnetomotive force, At
N = number of turns
I = current, A

Example 3.5

Problem:

Calculate the ampere-turns for a coil with 2,000 turns and a 5 Ma current.

Solution:

Use Equation (3.6) and substitute $N = 2{,}000$ and $I = 5 \times 10^{-3}$ A.

$$NI = 2{,}000(5 \times 10^{-3}) = 10 \text{ At}$$

The unit of *intensity* of magnetizing force per unit of length is designated as H and is sometimes expressed as Gilberts per centimeter of length. Expressed as an equation,

$$H = \frac{NI}{L} \tag{3.5}$$

where

H = magnetic field intensity, ampere-turns per meter (At/m)
NI = ampere-turns, At
L = length between poles of the coil, m

Note: Equation (3.5) is for a solenoid. H is the intensity of an air core. With an iron core, H is the intensity through the entire core and L is the length or distance between poles of the iron core.

Properties of Magnetic Materials

In this section, we discuss two important properties of magnetic materials: permeability and hysteresis.

Permeability

When the core of an electromagnet is made of annealed sheet steel it produces a stronger magnet than if a cast iron core is used. This is the case because annealed sheet steel is more readily acted upon by the magnetizing force of the coil than is the hard cast iron. Simply put, soft sheet steel is said to have greater *permeability* because of the greater ease with which magnetic lines are established in it.

Recall that permeability is the relative ease with which a substance conducts magnetic lines of force. The permeability of air is arbitrarily set at 1. The permeability of other substances is the ratio of their ability to conduct magnetic lines compared to that of air. The permeability of nonmagnetic materials, such as aluminum, copper, wood, and brass, is essentially unity, or the same as for air.

Important Note: The permeability of magnetic materials varies with the degree of magnetization, being smaller for high values of flux density. *Reluctance*—which is analogous to resistance is the opposition to the production of flux in a material—is inversely proportional to permeability. Iron has high permeability and, therefore, low reluctance. Air has low permeability and hence high reluctance.

Hysteresis

When the current in a coil of wire reverses thousands of times per second, a considerable loss of energy can occur. This loss of energy is caused by *hysteresis*. Hysteresis means "a lagging behind"; that is, the magnetic flux in an iron core lags behind the increases or decreases of the magnetizing force. The simplest method of illustrating the property of hysteresis is by graphical means such as the hysteresis loop shown in Figure 3.21.

The hysteresis loop (Figure 3.21) is a series of curves that show the characteristics of a magnetic material. Opposite directions of current result are in the opposite

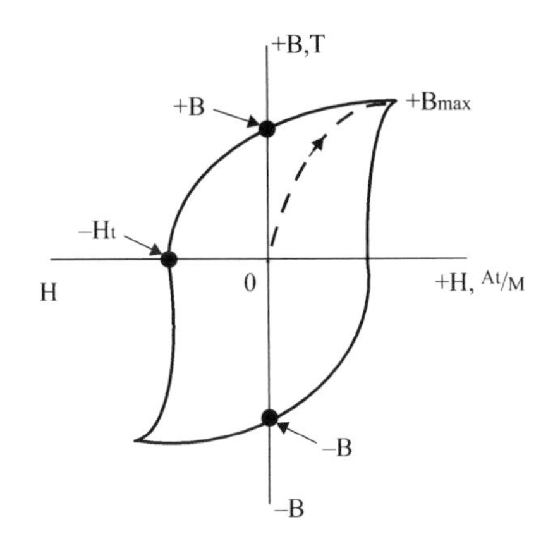

FIGURE 3.21 Hysteresis loop.

directions of +H and −H for field intensity. Similarly, opposite polarities are shown for flux density as +B and −B. The current starts at the center 0 (zero) when the material is unmagnetized. Positive H values increase B to saturation at $+B_{max}$. Next H decreases to zero, but B drops to the value of B, because of hysteresis. The current that produced the original magnetization now is reversed so that H becomes negative. B drops to zero and continues to $-B_{max}$. As the −H values decrease, B is reduced to −B, when H is zero. Now with a positive swing of current, H becomes positive, producing saturation at $+B_{max}$ again. The hysteresis loop is now completed. The curve doesn't return to zero at the center because of hysteresis.

Electromagnets

An *electromagnet* is composed of a coil of wire wound around a core that is normally soft iron, because of its high permeability and low hysteresis. When direct current flows through the coil, the core will become magnetized with the same polarity that the coil would have without the core. If the current is reversed, the polarity of the coil and core is reversed.

The electromagnet is of great importance in electricity simply because the magnetism can be "turned on" or "turned off" at will. The starter solenoid (an electromagnet) in automobiles and power boats is a good example. In an automobile or boat, an electromagnet is part of a relay that connects the battery to the induction coil, which generates the very high voltage needed to start the engine. The starter solenoid isolates this high voltage from the ignition switch. When no current flows in the coil, it is an "air core," but when the coil is energized, a movable soft iron core does two things. First, the magnetic flux is increased because the soft iron core is more permeable than the air core. Second, the flux is more highly concentrated. All this concentration of magnetic lines of force in the soft iron core results in a very good magnet when current flows in the coil. But soft iron loses its magnetism quickly when the current is shut off. The effect of the soft iron is, of course, the same whether it is movable, as in some solenoids, or permanently installed in the coil. An electromagnet, then consists basically of a coil and a core; it becomes a magnet when current flows through the coil.

The ability to control the action of magnetic force makes an electromagnet very useful in many circuit applications. Many of the applications of electromagnets are discussed throughout this manual.

ONE MORE WORD

This chapter has placed a building block in the foundation of understanding the science of Electric Vehicles (EVs). Although the material presented to this point is not designed to qualify anyone as an electrical engineer, electrician, or vehicle dynamics engineer, it is intended to set the stage for forthcoming material in the book, such as the next chapter dealing with vehicle motion or vehicle dynamics.

4 Battery-Supplied Electricity

Battery-supplied direct current (DC) electricity has many applications and is widely used in household, commercial, and industrial operations—also commonly used for starting, lighting, and ignition of vehicles. Beyond their applications in vehicles other somewhat well-known applications of battery-supplied systems include providing electrical energy in industrial vehicles and emergency diesel generators, material handling equipment (forklifts), portable electric/electronic equipment, backup emergency power for light-packs, for hazard warning signal lights and flashlights, and as standby power supplies or uninterruptible power supplies (UPS) for computer systems. In some instances, they are used as the only source of power, while in others (as mentioned above) they are used as a secondary or standby power supply. In renewable energy applications, batteries are used to store electrical energy. A battery pack stores electricity produced by a solar electric system. Today batteries commonly are used in specialized applications such as plug-in **battery electric vehicle (BEV)** and **hybrid electric vehicle (HEV)** rechargeable energy storage systems. Batteries are often used in wind-hybrid systems to store excess wind energy and then provide supplementary energy when the wind cannot generate sufficient power to meet the electric load. Batteries are also used in road-side solar-charging systems; that is, solar energy is used to charge a battery pack that supplies electrical power to some type of electrical signaling device, navigation buoys, or other remote application (e.g., emergency telephone, etc.) where it would be impractical and costly to run miles of electrical supply cable.

Note that in this book our main focus is on **electric vehicle battery (EVB**, also known as **a traction battery)** systems that are rechargeable and used to power the electric motors of BEVs or HEVs. These EVBs are typically **lithium-ion batteries**; they are specifically designed for high electric charge (aka energy) capacity. EVBs are discussed in detail in this chapter; however, this is a science book so to properly introduce any type of battery it is important to begin with battery basics and then advance, step-by-step, into a discussion of the presently available EVBs.

BASIC BATTERY TERMINOLOGY

- A voltaic cell is a combination of materials used to convert chemical energy into electrical energy.
- A battery is a group of two or more connected voltaic cells.
- An electrode is a metallic compound, or metal, which has an abundance of electrons (negative electrode) or an abundance of positive charges (positive electrode).
- An electrolyte is a solution that is capable of conducting an electric current.

DOI: 10.1201/9781003332992-4

- Specific gravity is defined as the ratio comparing the weight of any liquid to the weight of an equal volume of water.
- An ampere-hour is defined as a current of 1 ampere flowing for 1 hour.
- An external circuit is used to conduct the flow of electrons between the electrodes of the voltaic cell and usually includes a load.

BATTERY CHARACTERISTICS

Batteries are generally classified by their various characteristics. Parameters such as internal resistance, specific gravity, capacity, shelf life, "c" rate, MPV, gravimetric energy density, volumetric energy density, constant voltage charges, constant-current charge, and specific power are used to describe and classify batteries by operation and type.

Regarding **internal resistance**, it is important to keep in mind that a battery is a DC voltage generator. As such, the battery has internal resistance or equivalent series resistance (ESR). In a chemical cell, the resistance of the electrolyte between the electrodes is responsible for most of the cell's internal resistance. Because any current in the battery must flow through the internal resistance, this resistance is in series with the generated voltage. With no current, the voltage drop across the resistance is zero so that the full generated voltage develops across the output terminals. This is the open-circuit voltage, or no-load voltage. If a load resistance is connected across the battery, the load resistance is in series with internal resistance. When current flows in this circuit, the internal voltage drop decreases the terminal voltage of the battery.

The ratio of the weight of a certain volume of liquid to the weight of the same volume of water is called the **specific gravity** of the liquid. Pure sulfuric acid has a specific gravity of 1.835 since it weighs 1.835 times as much as water per unit volume. The specific gravity of a mixture of sulfuric acid and water varies with the strength of the solution from 1.000 to 1.830.

The specific gravity of the electrolyte solution in a lead-acid cell ranges from 1.210 to 1.300 for new, fully charged batteries. The higher the specific gravity, the less internal resistance of the cell and the higher the possible load current. As the cell discharges, the water formed dilutes the acid and the specific gravity gradually decreases to about 1.150, at which time the cell is considered to be fully discharged.

The specific gravity of the electrolyte is measured with a **hydrometer**, which has a compressible rubber bulb at the top, a glass barrel, and a rubber hose at the bottom of the barrel. In taking readings with a hydrometer, the decimal point is usually omitted. For example, a specific gravity of 1.260 is read simply as "twelve-sixty." A hydrometer reading of 1,210–1,300 indicates full charge; about 1,250 is half-charge; and 1,150–1,200 is complete discharge.

The **capacity** of a battery is measured in ampere-hours (Ah). The ampere-hour capacity is equal to the product of the current in amperes and the time in hours during which the battery is supplying this current and is defined as the amount of current that a battery can deliver for 1 hour before the battery voltage reaches the end-of-life point. In other words, 1 amp-hour is the equivalent of drawing 1 amp steadily for 1 hour, or 2 amps steadily for half an hour. A typical 12-V system may have 800 amp-hours of battery capacity. The battery can draw 100 amps for 8 hours if fully discharged and starting from a fully charged state. This is equivalent to 1,200 watts for 8 hours (power in watts = amps × volts). The ampere-hour capacity varies inversely with the discharge current. The size of a cell is determined generally by its ampere-hour capacity.

The capacity of a storage battery determines how long it will operate at a given discharge rate and depends upon many factors, the most important of which are as follows:

- The area of the plates in contact with the electrolyte
- The quantity and specific gravity of the electrolyte
- The type of separators
- The general condition of the battery (degree of sulfating, plates bucked, separators warped, sediment in bottom of cells, etc.)
- The final limiting voltage

The **shelf life** of a cell is that period of time during which the cell can be stored without losing more than approximately 10% of its original capacity. The loss of capacity of a stored cell is due primarily to the drying out of its electrolyte in a wet cell and to chemical actions that change the materials within the cell. The shelf life of a cell can be extended by keeping it in a cool, dry place.

The **"c" rate** is a current that is numerically equal to the Ah-rating of the cell. Charge and discharge currents are typically expressed in fractions of multiples of the c rate.

The **MPV (mid-point voltage)** is the nominal voltage of the cell and is the voltage that is measured when the battery has discharged 50% of its total energy.

The **gravimetric energy density** of a battery is a measure of how much energy a battery contains in comparison to its weight.

The **volumetric energy density** of a battery is a measure of how much energy a battery contains in comparison to its volume.

A **constant voltage charger** is a circuit that recharges a battery by sourcing only enough current to force the battery voltage to a fixed value.

A **constant-current charger** is a circuit that charges a battery by sourcing a fixed current into the battery, regardless of battery voltage.

In batteries, **specific power** usually refers to the power-to-weight ratio, measured in kilowatts per kilogram (generally, kW/kg).

VOLTAIC CELL

The simplest cell (an electrochemical device that transforms chemical energy into electrical energy) is known as a *voltaic* (or galvanic) cell (see Figure 4.1). It consists of a piece of carbon (C) and a piece of zinc (Zn) suspended in a jar that contains a solution of water (H_2O) and sulfuric acid (H_2SO_4).

Note: A simple cell consists of two strips, or **electrodes**, placed in a container that holds the **electrolyte**. A battery is formed when two or more cells are connected.

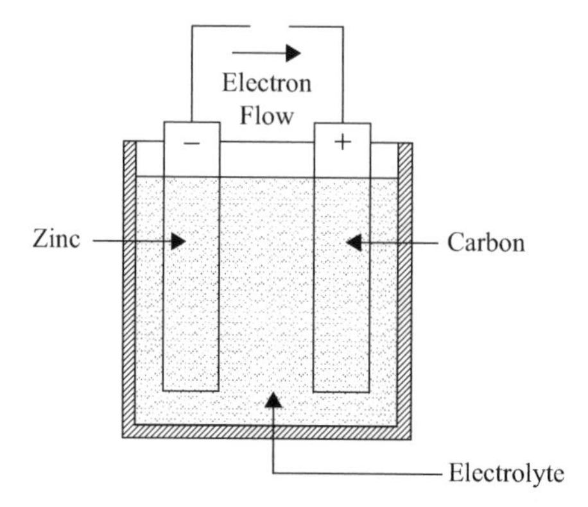

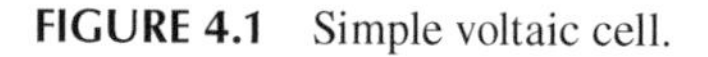
FIGURE 4.1 Simple voltaic cell.

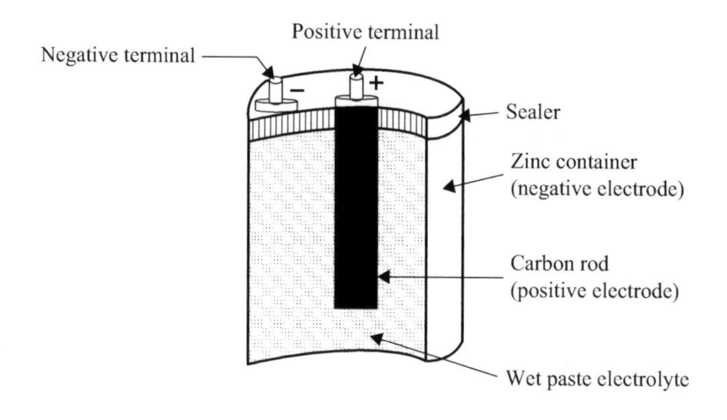

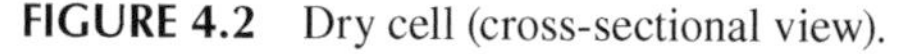
FIGURE 4.2 Dry cell (cross-sectional view).

The electrodes are the conductors by which the current leaves or returns to the electrolyte. In the simple cell described above, they are carbon and zinc strips placed in the electrolyte. Zinc contains an abundance of negatively charged atoms, while carbon has an abundance of positively charged atoms. When the plates of these materials are immersed in an electrolyte, chemical action between the two begins.

In the **dry cell** (see Figure 4.2), the electrodes are the carbon rod in the center and the zinc container in which the cell is assembled.

The electrolyte is the solution that acts upon the electrodes that are placed in it. The electrolyte may be a salt, an acid, or an alkaline solution. In the simple voltaic cell and in the automobile storage battery, the electrolyte is in a liquid form, while in the dry cell (see Figure 4.2), the electrolyte is a moist paste.

PRIMARY AND SECONDARY CELLS

Primary cells are normally those that cannot be recharged or returned to good condition after their voltage drops too low. Dry cells in flashlights and transistor radios

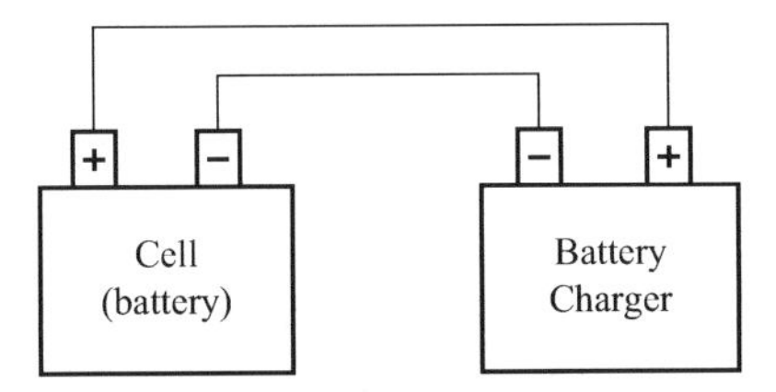

FIGURE 4.3 Hookup for charging a secondary cell with a battery charger.

are examples of primary cells. Some primary cells have been developed to the state where they can be recharged.

A *secondary cell* is one in which the electrodes and the electrolyte are altered by the chemical action that takes place when the cell delivers current. These cells are rechargeable. During recharging, the chemicals that provide electric energy are restored to their original condition. Recharging is accomplished by forcing an electric current through them in the opposite direction to that of discharge.

Connecting as shown in Figure 4.3 recharges a cell. Some battery chargers have a voltmeter and an ammeter that indicate the charging voltage and current.

The automobile storage battery is the most common example of the secondary cell.

BATTERY

As stated previously, a cell is an electrochemical device capable of supplying the energy that results from an internal chemical reaction to an external electrical circuit. In the simplest terms, batteries are made up of an anode, cathode, separator, electrolyte, and two current collectors (positive and negative). A *battery* consists of two or more cells placed in a common container. The cells are connected in series, in parallel, or in some combination of series and parallel, depending upon the amount of voltage and current required of the battery. The connection of cells in a battery is discussed in more detail later.

BATTERY OPERATION

The chemical reaction within a battery provides the voltage. This occurs when a conductor is connected externally to the electrodes of a cell, causing electrons to flow under the influence of a difference in potential across the electrodes from the zinc (negative) through the external conductor to the carbon (positive), returning within the solution to the zinc. After a short period of time, the zinc will begin to waste away because of the acid.

The voltage across the electrodes depends upon the materials from which the electrodes are made and the composition of the solution. The difference of potential between the carbon and zinc electrodes in a dilute solution of sulfuric acid and water is about 1.5 V.

The current that a primary cell may deliver depends upon the resistance of the entire circuit, including that of the cell itself. The internal resistance of the primary cell depends upon the size of the electrodes, the distance between them in the solution, and the resistance of the solution. The larger the electrodes and the closer together they are in solution (without touching), the lower the internal resistance of the primary cell and the more current it is capable of supplying to the load.

Note: When current flows through a cell, the zinc gradually dissolves in the solution and the acid is neutralized.

COMBINING CELLS

In many operations, battery-powered devices may require more electrical energy than one cell can provide. Various devices may require either a higher voltage or more current, and some cases both. Under such conditions it is necessary to combine, or interconnect, a sufficient number of cells to meet the higher requirements. Cells connected in series provide a higher voltage, while cells connected in parallel provide a higher current capacity. To provide adequate power when both voltage and current requirements are greater than the capacity of one cell, a combination series-parallel network of cells must be interconnected.

When cells are connected in **series** (see Figure 4.4), the total voltage across the battery of cells is equal to the sum of the voltage of each of the individual cells. In Figure 4.4 the four 1.5-V cells in series provide a total battery voltage of 6 V. When cells are placed in series, the positive terminal of one cell is connected to the negative terminal of the other cell. The positive electrode of the first cell and the negative electrode of the last cell then serve as the power takeoff terminals of the battery. The current flowing through such a battery of series cells is the same as from one cell because the same current flows through all the series cells.

To obtain a greater current, a battery has cells connected in **parallel** as shown in Figure 4.5. In this parallel connection, all the positive electrodes are connected to one line, and all negative electrodes are connected to the other. Any point on the positive side can serve as the positive terminal of the battery and any point on the negative side can be the negative terminal.

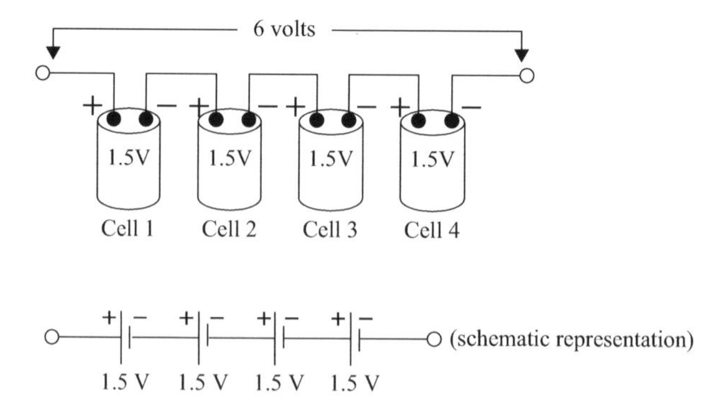

FIGURE 4.4 Cells in series.

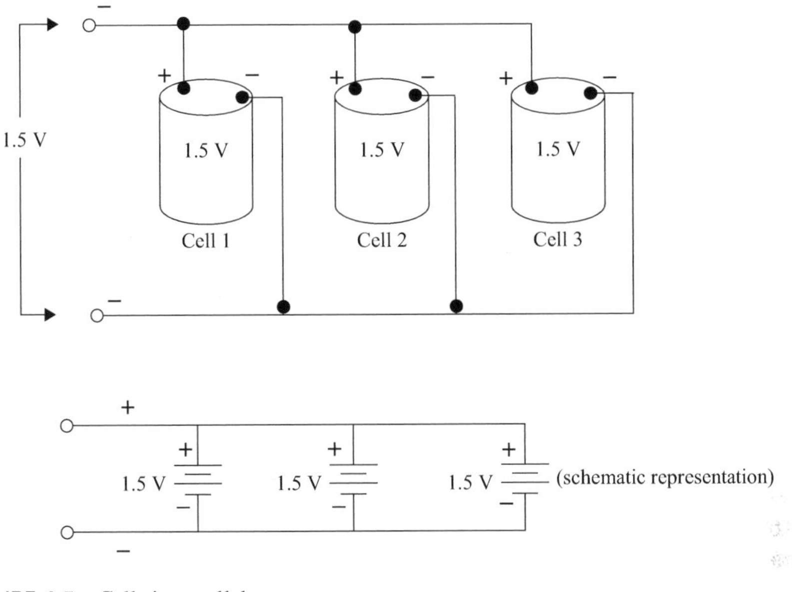

FIGURE 4.5 Cells in parallel.

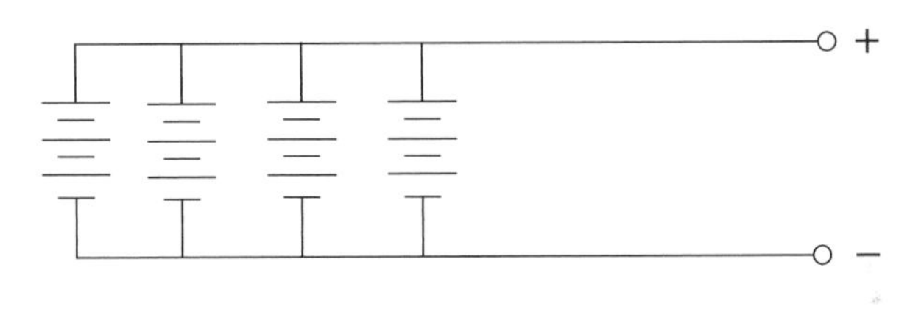

FIGURE 4.6 Series-parallel connected cells.

The total output voltage of a battery of three parallel cells is the same as that for a single cell (Figure 4.5), but the available current is three times that of one cell; that is, the current capacity has been increased.

Identical cells in parallel all supply equal parts of the current to the load. For example, of three different parallel cells producing a load current of 210 ma, each cell contributes 70 ma.

Figure 4.6 depicts a schematic of a **series-parallel** battery network supplying power to a load requiring both a voltage and current greater than one cell can provide. To provide the required increased voltage, groups of three 1.5-V cells are connected in series. To provide the required increased amperage, four series groups are connected in parallel.

TYPES OF BATTERIES

In the past 30 years, several different types of batteries have been developed. In this text we briefly discuss the dry cell battery, and those batteries that are currently used to store electrical energy in a reversible chemical reaction as applied to renewable energy production. The renewable energy source (solar, wind, or hydro) produces the energy, and the battery stores it for times of low or no renewable energy production. The types of batteries used in this application and discussed here include the lead-acid battery, alkaline cell, nickel-cadmium, mercury cell, nickel-metal hydride, lithium-ion, lithium-ion polymer batteries. Keep in mind that a battery does not create energy; instead, it stores energy. For most renewable energy applications, the preferred battery type is one that is a deep-cycle battery. A deep-cycle battery is designed to deliver a constant voltage as the battery discharges. A car starting battery, in contrast, is designed to deliver sporadic current spikes. Battery-driven vehicles, such as forklifts, golf carts, and floor sweepers, commonly use deep-cycle batteries. Deep-cycle battery can be charged with a lower current than regular batteries.

- **Dry cell**—The dry cell, or carbon-zinc cell, is so called because its electrolyte is not in a liquid state (however, the electrolyte is a moist paste). The dry cell battery is one of the oldest and most widely used commercial types of dry cell. The carbon, in the form of a rod that is placed in the center of the cell, is the positive terminal. The case of the cell is made of zinc, which is the negative terminal (see Figure 4.2). Between the carbon electrode and the zinc case is the electrolyte of a moist chemical paste-like mixture. The cell is sealed to prevent the liquid in the paste from evaporating. The voltage of a cell of this type is about 1.5 V.
- **Lead-acid battery**—The *lead-acid battery* is a secondary cell—commonly termed a storage battery or rechargeable—that stores chemical energy until it is released as electrical energy. The lead-acid battery differs from the primary cell type battery mainly in that it may be recharged, whereas most primary cells are not normally recharged. As the name implies, the lead-acid battery consists of a number of lead-acid cells immersed in a dilute solution of sulfuric acid. Each cell has two groups of lead plates; one set is the positive terminal and the other is the negative terminal. Active materials within the battery (lead plates and sulfuric acid electrolyte) react chemically to produce a flow of direct current whenever current consuming devices are connected to the battery terminal posts. This current is produced by chemical reaction between the active material of the plates (electrodes) and the electrolyte (sulfuric acid). This type of cell produces slightly more than 2 V. Most automobile batteries contain six cells connected in series so that the output voltage from the battery is slightly more than 12 V. Besides being rechargeable, the main advantage of the lead-acid storage battery over the dry cell battery is that the storage battery can supply current for a much longer time than the average dry cell. Lead-acid batteries can be designed to be high power and are inexpensive, safe, and reliable. Recycling infrastructure is in place for them. But low specific energy, poor cold temperature performance, and short calendar and cycle life are still impediments to their use. Advanced high-power, deep-cycle lead-acid batteries are being developed for Hybrid Electrical Vehicle (HEV) applications. However, lead-acid batteries are used for residential solar electric systems because of their low maintenance requirements and cost.

Safety Note: Whenever a lead-acid storage battery is charging, the chemical action produces dangerous hydrogen gas; thus, the charging operation should take place only in a well-ventilated area.

- **Alkaline cell**—The *alkaline cell* is a secondary cell that gets its name from its alkaline electrolyte—potassium hydroxide. Another type of battery, sometimes called the "alkaline battery," has a negative electrode of zinc and a positive electrode of manganese dioxide. It generates 1.5 V.
- **Nickel-cadmium cell**—The *nickel-cadmium cell*, or Ni-Cad cell, is the only dry battery that is a true storage battery with a reversible chemical reaction, allowing to be recharged many times. In the secondary nickel-cadmium dry cell, the electrolyte is potassium hydroxide, the negative electrode is nickel hydroxide, and the positive electrode is cadmium oxide. The operating voltage is 1.25 V. Because of its rugged characteristics (stands up well to shock, vibration, and temperature changes) and availability in a variety of shapes and sizes, it is ideally suited for use in powering portable communication equipment. Ni-Cad batteries are very expensive. Moreover, although nickel-cadmium batteries, used in many electronic consumer products, have higher specific energy and better life cycle than lead-acid batteries, they are of low efficiency (65%–80%) and do not deliver sufficient power and are not being considered for Hybrid Electric Vehicle (HEV) applications. Cadmium is a heavy metal that is toxic and is very expensive to dispose of, which reduces its desirability for use in hybrid application (NREL, 2009).
- **Mercury cell**—The *mercury cell* was developed as a result of space exploration activities, the development of small transceivers and miniaturized equipment where a power source of miniaturized size was needed. In addition to reduced size, the mercury cell has a good shelf life and is very rugged; it also produces a constant output voltage under different load conditions. There are two different types of mercury cells. One is a flat cell that is shaped like a button, while the other is a cylindrical cell that looks like a standard flashlight cell. The advantage of the button-type cell is that several of them can be stacked inside one container to form a battery. A cell produces 1.35 V.
- **Nickel-metal hydride**—Nickel-metal hydride batteries, used routinely in computer and medical equipment, offer reasonable specific energy and specific power capabilities. Their components are recyclable, but a recycling structure is not yet in place. Nickel-metal hydride batteries have a much longer life cycle than lead-acid batteries and are safe and abuse-tolerant. These batteries have been used successfully in the production of electric vehicles and recently in low-volume production of HEVs. The main challenges with nickel-metal hydride batteries are their high cost, high rate of self-discharge, very high gassing/waste consumption, and heat generation at high temperatures. The need is to control losses of hydrogen, and their low cell efficiency (may be as low as 50%, typically 60%–65%) (NREL, 2009).
- **Lithium-ion batteries (aka LIB or Li-ion)**—Lithium-ion batteries are rapidly penetrating into laptop and cellphone markets because of their high specific energy. They also have high specific power, high energy efficiency,

good high-temperature performance, and low self-discharge. Components of lithium-ion batteries can also be recycled. These characteristics make lithium-ion batteries suitable for HEV applications. However, to make them commercially viable for HEVs, further development is needed including improvement in calendar and cycle life, higher degree of cell and battery safety, abuse tolerance, and acceptable cost (NREL, 2009)—it can be said that all of these improvements are a work in progress and progress is being made on a constant basis.

- **Lithium-ion polymer batteries**—Lithium-ion polymer batteries with high specific energy (i.e., high energy per unit mass), initially developed for cellphone applications, also have the potential to provide high specific power for HEV applications. The other key characteristics of the lithium polymer are safety and good cycle life. The battery could be commercially viable if the cost is lowered and higher specific power batteries are developed (NREL, 2009).

ELECTRIC VEHICLE TERMINOLOGY

When a lithium-ion or lithium-ion polymer battery is used to power a BEV or HEV, in many cases, the terminology used or applied is the same. The key terms currently used are listed and defined in the following. Other terms not defined in the following are defined when introduced.

- **Alternating current (AC)**—a charge of electricity that periodically changes directions.
- **All-electric range (AER)**—the range an EV is able to reach safely using electricity.
- **All-electric vehicle (AEV)**—a BEV.
- **Combined charging system (CCS)**—a standard, fast charging system (50kW and up), often used with the SAE J1772 (J-Plug).
- **CHAdeMO**—the trademark of a standard, fast charging system (50kW and up).
- **Connector**—a device attached to a cable from an Electric Vehicle Supply Equipment (EVSE) that converts to an EV allowing it to charge.
- **Direct current (DC)**—a charge of electricity that flows in one direction and is the type of power that comes from a battery.
- **DC fast charging**—the fastest (high-powered) way to charge EVs—can charge a battery to 80% in 30 minutes, slows in order to not overheat the battery.
- **Energy density vs. power density**—energy density is measured in watt-hours per kilogram (Wh/kg) and is the amount of energy the battery can store with respect to its mass. Power density is measured in watts per kilogram (W/kg) and is the amount of power that can be generated by the battery with respect to its mass.
- **Extended-range electric vehicles (EREVs)**—have the ability to run on small internal combustion engines (called range extender) if the battery gets

low. Usually, the range extender powers a generator that transfers electricity to the batteries and motor.

- **Fuel cell electric vehicle (FCEV)**—uses compressed hydrogen gas for fuel.
- **Full hybrid electric vehicle (FGFV)**—combines a conventional internal combustion engine system with an electric propulsion system.
- **Level 1 (slow) charging (L1)**—every EV comes with a universally compatible L1 charge cable that plugs into any standard grounded 120-V outlet. The L1 charger power rating tops out at 2.4 kW, restoring about 5–8 miles per hour charge time, about 40 miles every 8 hours. Many drivers refer to the L1 charge cable as a trickle charger or emergency charger. The L1 charger will not keep up with long commutes or long drives anywhere.
- **Level 2 (fast) charging (L2)**—this charger runs at a higher input voltage, 240 V, and is usually a dedicated 240-V circuit in a driveway or garage that can be used with a J-plug connector. This is the most common charging system for residential use and can be found at commercial facilities. These chargers tend to top out at 12 kW, restoring up to 12–25 miles per hour charge, about 100 miles every 8 hours.
- **Level 3 (rapid) charging (L3)**—the fastest EV chargers available; it charges a battery to 80% in 30 minutes (using 480-V circuits), then slows to prevent overheating of the battery. Both CHAdeMO and SAE CCS connectors are used.
- **Molten salt battery**—uses molten salts as an electrolyte.
- **MPGe—million per gallon equivalent**—used to compare the fuel efficiency of EVs and internal combustion engines. It is determined by measuring how far an EV can travel on 33.7 kWh (the energy equivalent of one gallon of gas).
- **Range**—distance an EV can travel on pure electric power before the battery requires a recharge. **Range worry** occurs whenever a concern arises that the EV's battery power will run out before the destination is reached.
- **Regenerative braking**—used in EVs to transfer energy from the braking function to the battery for stored energy.
- **SAE J1772 (J-Plug)**—standard North American electrical connection for EVs—works with Level 1 and Level 2 systems.

ONE MORE WORD

This chapter has presented information on EVs and specifically on battery basics and batteries presently used in EVs. In the next chapter we add another building block in the foundation, that is, the electric-powered vehicle. The information provided deals with very basic electrical circuits and electrical circuit terminology.

REFERENCE

NREL (2009). *Ultracapacitors*. National Renewable Energy Laboratory. Accessed 01/29/10 @ http://www.nrel.gov/vehiclesandfuels/energystorage/ultracapacitors.html?print.

5 AC Theory

INTRODUCTION

So, why are we discussing AC theory in this chapter when we all know that EVs are driven by batteries and batteries deliver DC electricity to the prime movers (electric motors) of the EV vehicle? Well, that question is what this chapter is all about and about to answer.

First, it is magnetism, whether AC or DC is the result is not important when it comes to powering today's EVs.

Why?

Well, the truth be told the EV can be powered by either DC or AC electrical power.

Well, an EV has a rechargeable battery—and that batteries produce DC electricity, so where does the AC power come into play?

Ah! And now we have reached the gist, that is, the point of this segment of our discussion. So, we know that an electric vehicle usually has an on-board rechargeable electrical storage unit. Generally a lithium-ion battery that works as the source of power for an electric motor powers the electric vehicle into motion.

Motion! A key term when it comes to transportation. And motion is what travel is all about. The point being that one can't get from point A to point B to point C and beyond and then ultimately return (if so desired) without motion that is either human-accomplished physically by walking or bicycling or mechanically by driving or riding in some form of transportation—not a horse but a mechanically powered vehicle. Keep in mind that the source of power for an electric motor that powers (puts it into motion) is a battery.

Lithium-ion batteries in EVs are what powers the vehicle into motion—lithium-ion batteries fit the bill, so to speak, because they have high power density, high energy density, and long life as compared with others.

Okay, all the above information is great and informative but the question lingers: What motor does an electric vehicle use—is it an AC or DC motor?

As stated earlier either AC or DC electricity is used in today's EVs. Therefore, in a science sense it is important to have a very basic understanding of the AC just as we did with DC battery science.

BASIC AC THEORY AND PRINCIPLES

Because voltage is induced in a conductor when lines of force are cut, the amount of the induced electromotive force (emf) depends on the number of lines cut in unit time. To induce an emf of 1 V, a conductor must cut 100,000,000 lines of force per second. To obtain this great number of "cuttings," the conductor is formed into a loop and rotated on an axis at great speed (see Figure 5.1). The two sides of the loop become individual conductors in series, each side of the loop cutting lines of

DOI: 10.1201/9781003332992-5

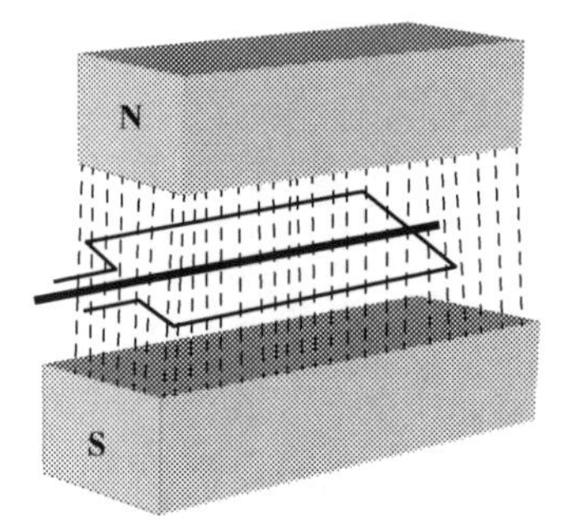

FIGURE 5.1 Loop rotating in magnetic field produces an AC voltage.

force and inducing twice the voltage that a single conductor would induce. In commercial generators, the number of "cuttings" and the resulting emf are increased by: (1) increasing the number of lines of force by using more magnets or stronger electromagnets, (2) using more conductors or loops, and (3) rotating the loops faster. (Note: both AC and DC generators are covered later.)

How an AC generator operates to produce an AC voltage and current is a basic concept today, taught in elementary and middle school science classes. Of course, we accept technological advances as commonplace today—we surf the internet, text our friends, watch cable television, use our cellphones, and take outer space flight as a given—and consider producing the electricity that makes all these technologies possible as our right. These technologies are most common to us today—we have them available to us so we simply use them—no big deal, right? Not worth thinking about. This point of view surely was not held initially—especially by those who broke ground in developing technology and electricity.

In groundbreaking years of electric technology development, the geniuses of the science of electricity (including George Simon Ohm) performed their technological breakthroughs in faltering steps. We tend to forget that those first faltering steps of scientific achievement in the field of electricity were performed with crude and, for the most part, homemade apparatus. (Sounds something like the more contemporary young garage-and-basement inventors who came up with the first basic user-friendly microcomputer and software packages, and hopefully the renewable energy innovations of tomorrow?)

Indeed, the innovators of electricity had to fabricate nearly all the laboratory equipment used in their experiments. At the time, the only convenient source of electrical energy available to these early scientists was the voltaic cell, invented some years earlier. Because of the fact that cells and batteries were the only sources of power available, some of the early electrical devices were designed to operate from **direct current**.

Thus, initially, direct current was used extensively. However, when the use of electricity became widespread, certain disadvantages in the use of direct current became apparent. In a direct current system the supply voltage must be generated at the level required by the load. To operate a 240-V lamp, for example, the generator must deliver 240 V. A 120-V lamp could not be operated from this generator by any convenient means. A resistor could be placed in series with the 120-V lamp to drop the extra 120 V, but the resistor would waste an amount of power equal to that consumed by the lamp.

Another disadvantage of direct current systems is the large amount of power lost due to the resistance of the transmission wires used to carry current from the generating station to the consumer. This loss could be greatly reduced by operating the transmission line at very high voltage and low current. This is not a practical solution in a DC system, however, since the load would also have to operate at high voltage. As a result of the difficulties encountered with direct current, practically all modern power distribution systems use **alternating current** (AC).

Unlike DC voltage, AC voltage can be stepped up or down by a device called a **transformer** (discussed later). Transformers permit the transmission lines to be operated at high voltage and low current for maximum efficiency. Then, at the consumer end, the voltage is stepped down to whatever value the load requires by using a transformer. Due to its inherent advantages and versatility, alternating current has replaced direct current in all but a few commercial power distribution systems.

Basic AC Generator

As shown in Figure 5.1, an AC voltage and current can be produced when a conductor loop rotates through a magnetic field and cuts lines of force to generate an induced AC voltage across its terminals. This describes the basic principle of operation of an alternating current generator, or **alternator**. An alternator converts mechanical energy into electrical energy. It does this by utilizing the principle of **electromagnetic induction**. The basic components of an alternator are an armature, about which many turns of conductor are wound, which rotates in a magnetic field, and some means of delivering the resulting alternating current to an external circuit. (Note: We cover generator construction in more detail; in this section, we concentrate on the theory of operation.)

Cycle

An AC voltage is one that continually changes in magnitude and periodically reverses in polarity (see Figure 5.2). The zero axis is a horizontal line across the center. The vertical variations on the voltage wave show the changes in magnitude. The voltages above the horizontal axis have positive (+) polarity, while voltages below the horizontal axis have negative (–) polarity.

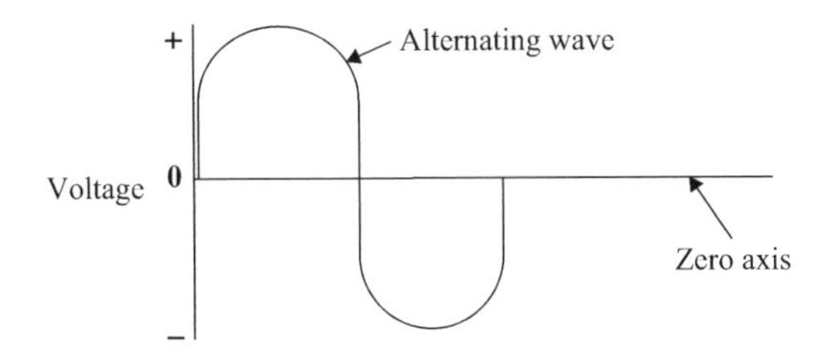

FIGURE 5.2 An AC voltage waveform.

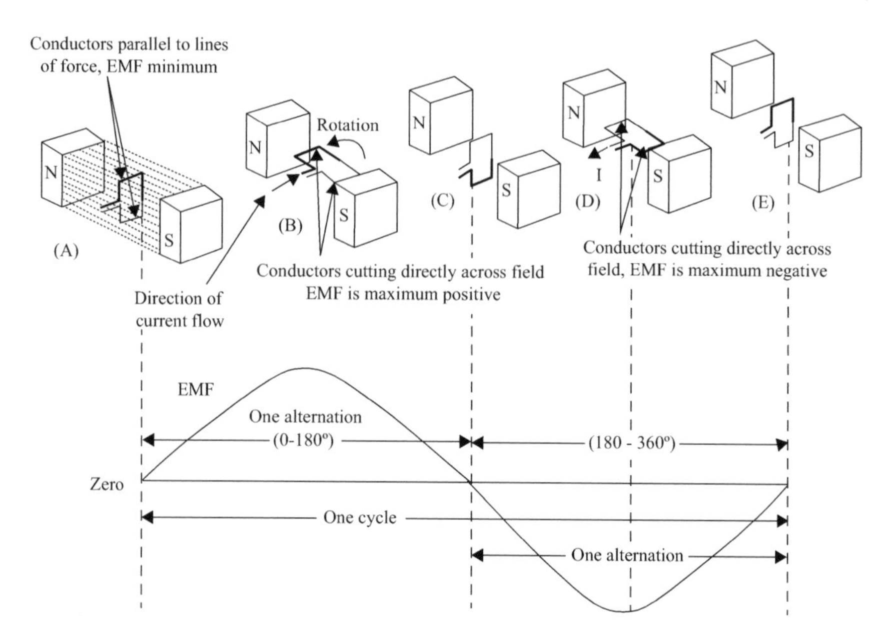

FIGURE 5.3 Basic alternating current generator.

Note: To bring the important points presented up to this point into finer focus and to expand the presentation, Figure 5.3 is provided with accompanying explanation.

Figure 5.3 shows a suspended loop of wire (conductor or armature) being rotated (moved) in a counterclockwise direction through the magnetic field between the poles of a permanent magnet. For ease of explanation, the loop has been divided into a thick and thin half. Notice that in part (A), the thick half is moving along (parallel to) the lines of force. Consequently, it is cutting none of these lines. The same is true of the thin half, moving in the opposite direction. Because the conductors are not cutting any lines of force, no emf is induced. As the loop rotates toward the position shown in part (B), it cuts more and more lines of force per second because it is cutting more directly across the field (lines of force) as it approaches the position shown in (B). At position (B), the induced voltage is greatest because the conductor is cutting directly across the field.

As the loop continues to be rotated toward the position shown in part (C), it cuts fewer and fewer lines of force per second. The induced voltage decreases from its peak value. Eventually, the loop is once again moving in a plane parallel to the magnetic field, and no voltage (zero voltage) is induced. The loop has now been rotated through half a circle (one alternation, or 180°). The sine curve shown in the lower part of Figure 5.3 shows the induced voltage at every instant of rotation of the loop. Notice that this curve contains 360°, or two alternations.

Important Point: Two complete alternations in a period of time is called a *cycle*.

In Figure 5.3, if the loop is rotated at a steady rate, and if the strength of the magnetic field is uniform, the number of cycles per second (cps), or **Hertz**, and the voltage will remain at fixed values. Continuous rotation will produce a series of sine wave voltage cycles, or, in other words, an AC voltage. In this way mechanical energy is converted into electrical energy.

Frequency, Period, and Wavelength

The *frequency* of an alternating voltage or current is the number of complete cycles occurring in each second of time. It is indicated by the symbol f and is expressed in hertz (Hz). One cycle per second equals 1 Hz. Thus 60 cycles per second (cps) equals 60 Hz. A frequency of 2 Hz (Figure 5.4b) is twice the frequency of 1 Hz (Figure 5.4a).

The amount of time for the completion of 1 cycle is the *period*. It is indicated by the symbol T for time and is expressed in seconds (s). Frequency and period are reciprocals of each other.

$$f = \frac{1}{T} \tag{5.1}$$

$$T = \frac{1}{f} \tag{5.2}$$

Important Point: The higher the frequency, the shorter the period.

The angle of 360° represents the time for 1 cycle, or the period T. So we can show the horizontal axis of the sine wave in units of either electrical degrees or seconds (see Figure 5.5).

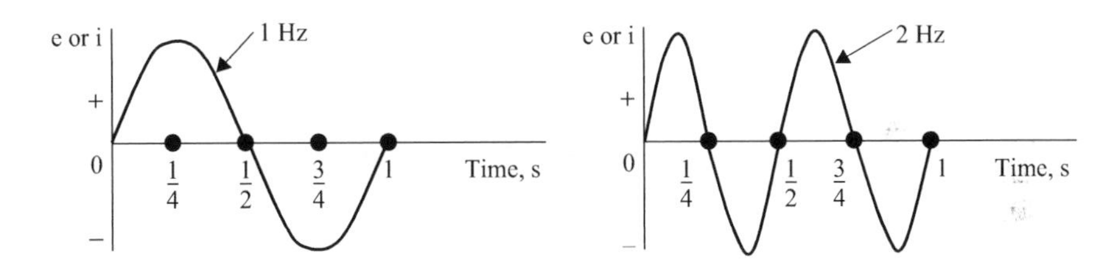

FIGURE 5.4 Comparison of frequencies.

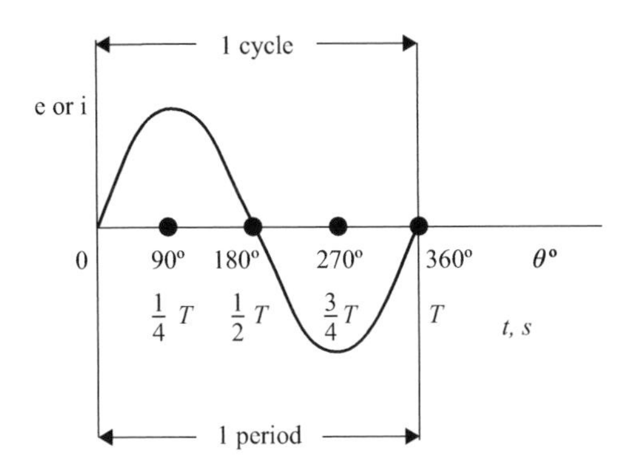

FIGURE 5.5 Relationship between electrical degrees and time.

The *wavelength* is the length of one complete wave or cycle. It depends upon the frequency of the periodic variation and its velocity of transmission. It is indicated by the symbol λ (Greek lowercase lambda). Expressed as a formula:

$$\lambda = \frac{\text{velocity}}{\text{frequency}} \tag{5.3}$$

CHARACTERISTIC VALUES OF AC VOLTAGE AND CURRENT

Because an AC sine wave voltage or current has many instantaneous values throughout the cycle, it is convenient to specify magnitudes for comparing one wave with another. The peak, average, or root-mean-square (rms) value can be specified (see Figure 5.6). These values apply to current or voltage.

PEAK AMPLITUDE

One of the most frequently measured characteristics of a sine wave is its amplitude. Unlike DC measurement, the amount of alternating current or voltage present in a circuit can be measured in various ways. In one method of measurement, the maximum amplitude of either the positive or the negative alternation is measured. The value of current or voltage obtained is called the *peak voltage* or the *peak current*. To measure the peak value of current or voltage, an oscilloscope must be used. The peak value is illustrated in Figure 5.6.

PEAK-TO-PEAK AMPLITUDE

A second method of indicating the amplitude of a sine wave consists of determining the total voltage or current between the positive and negative peaks. This value of current or voltage is called the *peak-to-peak value* (see Figure 5.6). Because both alternations of a pure sine wave are identical, the peak-to-peak value is twice the

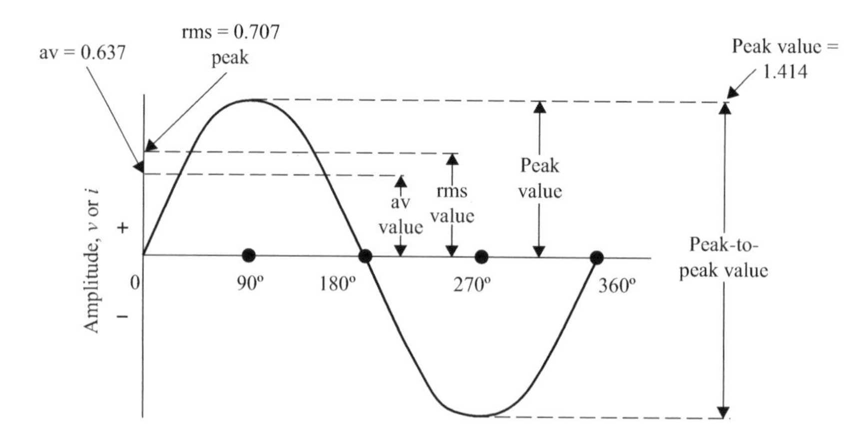

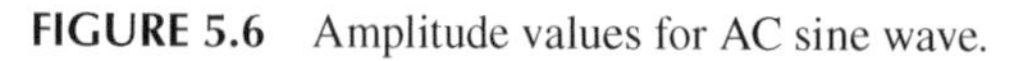
FIGURE 5.6 Amplitude values for AC sine wave.

peak value. Peak-to-peak voltage is usually measured with an oscilloscope, although some voltmeters have a special scale calibrated in peak-to-peak volts.

Instantaneous Amplitude

The *instantaneous value* of a sine wave of voltage for any angle of rotation is expressed by the formula:

$$e = E_m \times \sin\theta \tag{5.4}$$

where
- e = the instantaneous voltage
- E_m = the maximum or peak voltage
- $\sin\theta$ = the sine of angle at which e is desired

Similarly the equation for the instantaneous value of a sine wave of current would be:

$$i = I_m \times \sin\theta \tag{5.5}$$

where
- i = the instantaneous current
- I_m = the maximum or peak current
- $\sin\theta$ = the sine of the angle at which i is desired

Note: The instantaneous value of voltage constantly changes as the armature of an alternator moves through a complete rotation. Because current varies directly with voltage, according to Ohm's Law, the instantaneous changes in current also result in a sine wave whose positive and negative peaks and intermediate values can be plotted exactly as we plotted the voltage sine wave. However, instantaneous values are not useful in solving most AC problems, so an **effective** value is used.

Effective or RMS Value

The *effective value* of an AC voltage or current of sine waveform is defined in terms of an equivalent heating effect of a direct current. Heating effect is independent of the direction of current flow.

Important Point: Because all instantaneous values of induced voltage are somewhere between zero and E_M (maximum, or peak voltage), the effective value of a sine wave voltage or current must be greater than zero and less than E_M (the maximum, or peak voltage).

The alternating current of sine waveform having a maximum value of 14.14 amps produces the same amount of heat in a circuit having a resistance of 1 ohm as a direct current of 10 amps. Because this is true, we can work out a constant value for converting any peak value to a corresponding effective value. This constant is represented by X in the simple equation below. Solve for X to three decimal places.

$$14.14X = 10$$

$$X = 0.707$$

The effective value is also called the *root-mean-square (rms)* value because it is the square root of the average of the squared values between zero and maximum. The effective value of an AC current is stated in terms of an equivalent DC current. The phenomenon used as the standard comparison is the heating effect of the current.

Important Point: Anytime an AC voltage or current is stated without any qualifications, it is assumed to be an effective value.

In many instances it is necessary to convert from effective to peak value or vice versa using a standard equation. Figure 5.6 shows that the peak value of a sine wave is 1.414 times the effective value; therefore, the equation we use is:

$$E_m = E \times 1.414 \tag{5.6}$$

where

E_m = maximum or peak voltage
E = effective or RMS voltage

and

$$I_m = I \times 1.414 \tag{5.7}$$

where

I_m = maximum or peak current
I = effective or RMS current

On occasion, it is necessary to convert a peak value of current or voltage to an effective value. This is accomplished by using the following equations:

$$E = E_m \times 0.707 \tag{5.8}$$

where

E = effective voltage
E_m = the maximum or peak voltage

$$I = I_m \times 0.707 \tag{5.9}$$

where

I = the effective current
I_m = the maximum or peak current

Average Value

Because the positive alternation is identical to the negative alternation, the *average value* of a complete cycle of a sine wave is zero. In certain types of circuits however, it is necessary to compute the average value of one alternation. Figure 5.6 shows that the average value of a sine wave is 0.637 × peak value and therefore:

$$\text{Average Value} = 0.637 \times \text{peak value} \tag{5.10}$$

or

$$E_{avg} = E_m \times 0.637$$

where
E_{avg} = the average voltage of one alternation
E_m = the maximum or peak voltage

Similarly

$$I_{avg} = I_m \times 0.637 \tag{5.11}$$

where
I_{avg} = the average current in one alternation
I_m = the maximum or peak current

Table 5.1 lists the various values of sine wave amplitude used to multiply in the conversion of AC sine wave voltage and current.

Resistance in AC Circuits

If a sine wave of voltage is applied to a resistance, the resulting current will also be a sine wave. This follows Ohm's Law that states that the current is directly proportional to the applied voltage. Figure 5.1 shows a sine wave of voltage and the resulting sine wave of current superimposed on the same time axis. Notice that as the voltage increases in a positive direction the current increases along with it. When the voltage reverses direction, the current reverses direction. At all times the voltage and current

TABLE 5.1
AC Sine Wave Conversion Table

Multiply the Value	By	To Get the Value
Peak	2	Peak-to-peak
Peak-to-peak	0.5	Peak
Peak	0.637	Average
Average	1.637	Peak
Peak	0.707	RMS (effective)
RMS (effective)	1.414	Peak
Average	1.110	RMS (effective)
RMS (effective)	0.901	Average

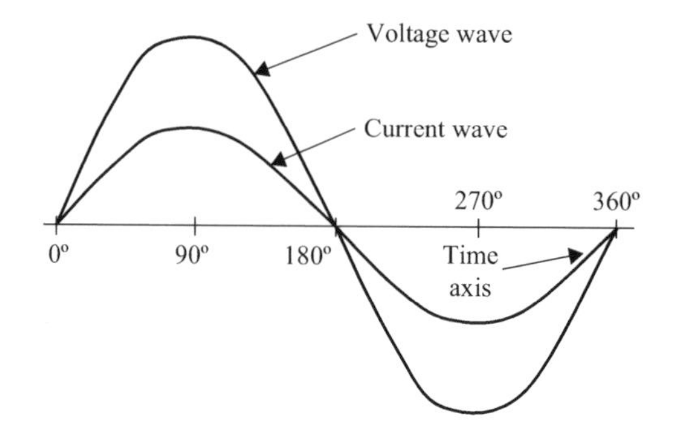

FIGURE 5.7 Voltage and current waves in phase.

pass through the same relative parts of their respective cycles at the same time. When two waves, such as those shown in Figure 5.7, are precisely in step with one another they are said to be **in phase**. To be in phase, the two waves must go through their maximum and minimum points at the same time and in the same direction.

In some circuits, several sine waves can be in phase with each other. Thus, it is possible to have two or more voltage drops in phase with each other and also in phase with the circuit current.

Note: It is important to remember that Ohm's Law for DC circuits is applicable to AC circuits with resistance only.

Voltage waves are not always in phase. For example, Figure 5.8 shows a voltage wave E_1 considered to start at 0° (time 1). As voltage wave E_1 reaches its positive peak, a second voltage wave E_2 starts to rise (time 2). Because these waves do not go through their maximum and minimum points at the same instant of time, a **phase difference** exists between the two waves. The two waves are said to be out of phase. For the two waves in Figure 5.8, this phase difference is 90°.

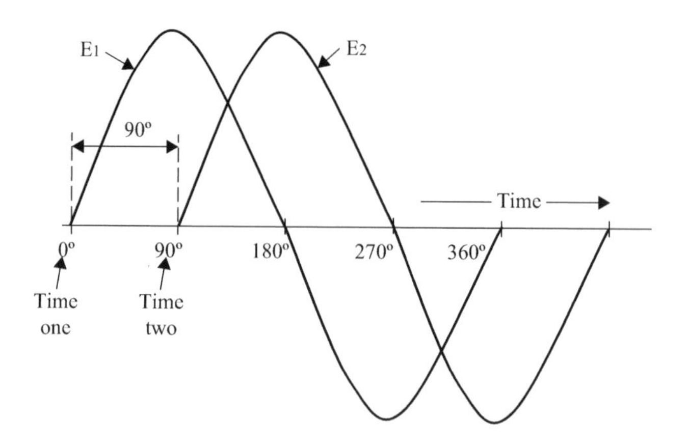

FIGURE 5.8 Voltage waves 90° out of phase.

Phase Relationships

In the preceding section we discussed the important concepts of **in phase** and **phase difference**. Another important phase concept is phase angle. The *phase angle* between two waveforms of the same frequency is the angular difference at a given instant of time. As an example, the phase angle between waves B and A (see Figure 5.9) is 90°. Take the instant of time at 90°. The horizontal axis is shown in angular units of time. Wave B starts at maximum value and reduces to zero value at 90°, while wave A starts at zero and increases to maximum value at 90°. Wave B reaches its maximum value 90° ahead of wave A, so wave B **leads** wave A by 90° (and wave A **lags** wave B by 90°). This 90° phase angle between waves B and A is maintained throughout the complete cycle and all successive cycles. At any instant of time, wave B has the value that wave A will have 90° later. Wave B is a cosine wave because it is displaced 90° from wave A, which is a sine wave.

Important Point: The amount by which one wave leads or lags another is measured in degrees.

To compare phase angles or phases of alternating voltages or currents, it is more convenient to use vector diagrams corresponding to the voltage and current waveforms. A *vector* is a straight line used to denote the magnitude and direction of a given quantity. Magnitude is denoted by the length of the line drawn to scale, and the direction is indicated by the arrow at one end of the line, together with the angle that the vector makes with a horizontal reference vector.

Note: In electricity, since different directions really represent **time** expressed as a phase relationship, an electrical vector is called a **phasor**. In an AC circuit containing only resistance, the voltage and current occur at the **same time**, or are in phase. To indicate this condition by means of phasors all that is necessary is to draw the phasors for the voltage and current in the same direction. The value of each is indicated by the **length** of the phasor.

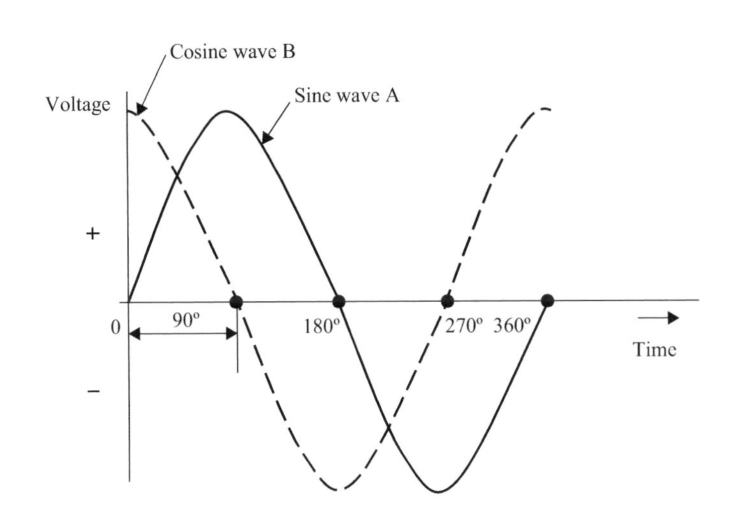

FIGURE 5.9 Wave B leads wave A by a phase angle of 90°.

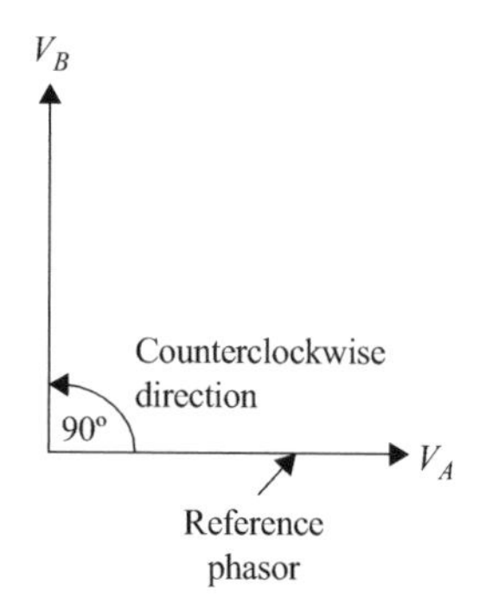

FIGURE 5.10 Phasor diagram.

A vector, or phasor, diagram is shown in Figure 5.10 where vector V_B is vertical to show the phase angle of 90° with respect to vector V_A, which is the reference. Since lead angles are shown in the counterclockwise direction from the reference vector, V_B leads V_A by 90°.

Inductance

To this point we have learned the following key points about magnetic fields:

- A field of force exists around a wire carrying a current.
- This field has the form of concentric circles around the wire, in planes perpendicular to the wire, and with the wire at the center of the circles.
- The strength of the field depends on the current. Large currents produce large fields; small currents produce small fields.
- When lines of force cut across a conductor, a voltage is induced in the conductor.

To this point we have studied circuits that have been **resistive** (i.e., resistors presented the only opposition to current flow). Two other phenomena—inductance and capacitance—exist in DC circuits to some extent, but they are major players in AC circuits. Both inductance and capacitance present a kind of opposition to current flow that is called **reactance**, which we will cover later. Before we examine reactance, however, we must first study inductance and capacitance.

Inductance: What Is It?

Inductance is the characteristic of an electrical circuit that makes itself evident by opposing the starting, stopping, or changing of current flow. A simple analogy can be used to explain inductance. We are all familiar with how difficult it is to push a heavy load (a cart full of heavy materials, etc.). It takes more work to start to move the load than it does to keep it moving. This is because the load possesses the property of **inertia**. Inertia is the characteristic of mass that opposes a **change** in velocity.

Therefore, inertia can hinder us in some ways and help us in others. Inductance exhibits the same effect on current in an electric circuit as inertia does on velocity of a mechanical object. The effects of inductance are sometimes desirable—sometimes undesirable.

Important Point: Simply put, **inductance** is the characteristic of an electrical conductor that opposes a **change** in current flow.

Because inductance is the property of an electric circuit that opposes any **change** in the current through that circuit, if the current increases, a self-induced voltage opposes this change and delays the increase. On the other hand, if the current decreases, a self-induced voltage tends to aid (or prolong) the current flow, delaying the decrease. Thus, current can neither increase nor decrease as fast in an inductive circuit as it can in a purely resistive circuit.

In AC circuits, this effect becomes very important because it affects the **phase** relationships between voltage and current. Earlier we learned that voltages (or currents) can be out of phase if they are induced in separate armatures of an alternator. In that case, the voltage and current generated by each armature were in phase. When inductance is a factor in a circuit, the voltage and current generated by the **same** armature are out of phase. We shall examine these phase relationships later. Our objective now is to understand the nature and effects of inductance in an electric circuit.

Unit of Inductance

The unit for measuring inductance, L, is the *Henry* (named for the American physicist, Joseph Henry), abbreviated *h* and normally written in lower case, henry. Figure 5.11 shows the schematic symbol for an inductor. An inductor has an inductance of 1 henry if an emf of 1 V is induced in the inductor when the current through the inductor is changing at the rate of 1 ampere per second. The relation between the induced voltage, inductance, and rate of change of current with respect to time is stated mathematically as

$$E = L\frac{\Delta I}{\Delta t} \tag{5.12}$$

where

E = the induced emf in volts

L = the inductance in henry

ΔI = is the change in amperes occurring in Δt seconds

Note: The symbol Δ (Delta) means "a change in …."

FIGURE 5.11 Schematic symbol for an inductor.

The henry is a large unit of inductance and is used with relatively large inductors. The unit employed with small inductors is the millihenry (mh). For still smaller inductors the unit of inductance is the microhenry (μh).

Self-Inductance

As previously explained, current flow in a conductor always produces a magnetic field surrounding, or linking with, the conductor. When the current changes, the magnetic field changes, and an emf is induced in the conductor. This emf is called a *self-induced emf* because it is induced in the conductor carrying the current.

Note: Even a perfectly straight length of conductor has some inductance.

The direction of the induced emf has a definite relation to the direction in which the field that induces the emf varies. When the current in a circuit is increasing, the flux linking with the circuit is increasing. This flux cuts across the conductor and induces an emf in the conductor in such a direction to oppose the increase in current and flux. This emf is sometimes referred to as **counterelectromotive force** (cemf). The two terms are used synonymously throughout this manual. Likewise, when the current is decreasing, an emf is induced in the opposite direction and opposes the decrease in current.

Important Point: The effects just described are summarized by **Lenz's Law**, which states that the induced emf in any circuit is always in a direction opposed to the effect that produced it.

Shaping a conductor so that the electromagnetic field around each portion of the conductor cuts across some other portion of the same conductor increases the inductance. This is shown in its simplest form in Figure 5.12a. A conductor is looped so that two portions of the conductor lie adjacent and parallel to one another. These portions are labeled Conductor 1 and Conductor 2. When the switch is closed, electron flow through the conductor establishes a typical concentric field around **all** portions of the conductor. The field is shown in a single plane (for simplicity) that is perpendicular to both conductors. Although the field originates simultaneously in both conductors it is considered as originating in Conductor 1 and its effect on Conductor 2 will be noted. With increasing current, the field expands outward, cutting across a portion of Conductor 2. The resultant induced emf in Conductor 2 is shown by the dashed arrow. Note that it is in **opposition** to the battery current and voltage, according to Lenz's Law.

In Figure 5.12b, the same section of Conductor 2 is shown, but with the switch opened and the flux collapsing.

Important Point: From Figure 5.12, the important point to note is that the voltage of self-induction opposes both **changes** in current. It delays the initial buildup of current by opposing the battery voltage and delays the breakdown of current by exerting an induced voltage in the same direction that the battery voltage acted.

Four major factors affect the self-inductance of a conductor, or circuit.

1. **Number of turns**—Inductance depends on the number of wire turns. Wind more turns to increase inductance. Take turns off to decrease the inductance. Figure 5.13 compares the inductance of two coils made with different numbers of turns.

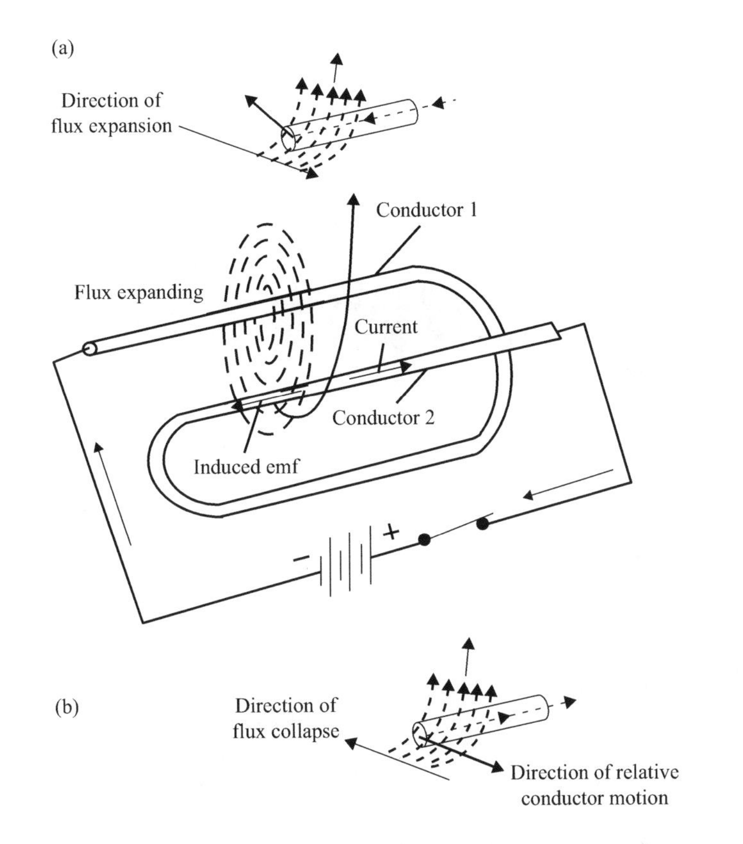

FIGURE 5.12 (a and b) Self-inductance.

FIGURE 5.13 (a) Few turns, low inductance; (b) more turns, higher inductance.

FIGURE 5.14 (a) Wide spacing between turns, low inductance; (b) close spacing between turns, higher inductance.

2. **Spacing between turns**—Inductance depends on the spacing between turns, or the inductor's length. Figure 5.14 shows two inductors with the same number of turns. The first inductor's turns have a wide spacing. The second inductor's turns are close together. The second coil, though shorter, has a larger inductance value because of its close spacing between turns.

FIGURE 5.15 (a) Small diameter, low inductance; (b) larger diameter, higher inductance.

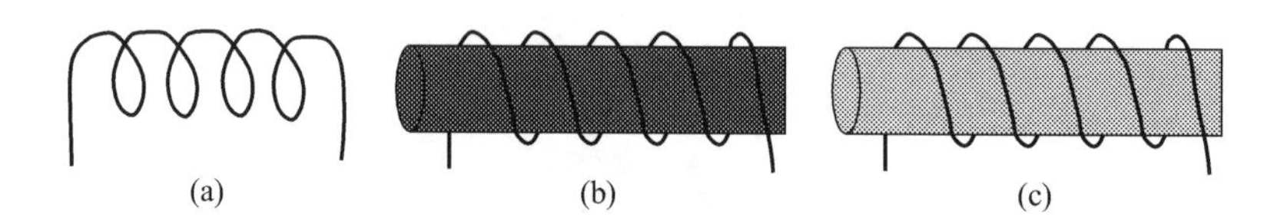

FIGURE 5.16 (a) air core, low inductance; (b) powdered iron core, higher inductance; (c) soft iron core, highest inductance.

3. **Coil diameter**—Coil diameter, or cross-sectional area, is highlighted in Figure 5.15. The larger-diameter inductor has more inductance. Both coils shown have the same number of turns, and the spacing between turns is the same. The first inductor has a small diameter and the second one has a larger diameter. The second inductor has more inductance than the first one.
4. **Type of core material—Permeability**, as pointed out earlier, is a measure of how easily a magnetic field goes through a material. Permeability also tells us how much stronger the magnetic field will be with the material inside the coil. Figure 5.16 shows three identical coils. One has an air core; one has a powdered iron core in the center and the other has a soft iron core. This figure illustrates the effects of core material on inductance. The inductance of a coil is affected by the magnitude of current when the core is a magnetic material. When the core is air, the inductance is independent of the current.

Key Point: The inductance of a coil increases very rapidly as the number of turns is increased. It also increases as the coil is made shorter, the cross-sectional area is made larger, or the permeability of the core is increased.

Growth and Decay of Current in an RL Series Circuit

If a battery is connected across a pure inductance, the current builds up to its final value at a rate that is determined by the battery voltage and the internal resistance of the battery. The current buildup is gradual because of the counter emf (cemf) generated by the self-inductance of the coil. When the current starts to flow, the magnetic lines of force move out, cut the turns of wire on the inductor, and build up a cemf that opposes the emf of the battery. This opposition causes a delay in the time it takes the current to build up to steady value. When the battery is disconnected, the lines of force collapse, again cutting the turns of the inductor and building up an emf that tends to prolong the current flow.

Although the analogy is not exact, electrical inductance is somewhat like mechanical inertia. A boat begins to move on the surface of water at the instant a constant force is applied to it. At this instant its rate of change of speed (acceleration) is greatest, and all the applied force is used to overcome the inertia of the boat. After a while the speed of the boat increases (its acceleration decreases) and the applied force is used up in overcoming the friction of the water against the hull. As the speed levels off and the acceleration becomes zero, the applied force equals the opposing friction force at this speed and the inertia effect disappears. In the case of inductance, it is electrical inertia that must be overcome.

Mutual Inductance

When the current in a conductor or coil changes, the varying flux can cut across any other conductor or coil located nearby, thus inducing voltages in both. A varying current in L_1, therefore, induces voltage across L_1 and across L_2 (see Figure 5.17; see Figure 5.18 for the schematic symbol for two coils with mutual inductance). When the induced voltage e_{L2} produces current in L_2, its varying magnetic field induces voltage in L_1. Hence, the two coils L_1 and L_2 have *mutual inductance* because current change in one coil can induce voltage in the other. The unit of mutual inductance is the henry, and the symbol is L_M. Two coils have L_M of 1 H when a current change of 1 A/s in one coil induces 1 E in the other coil.

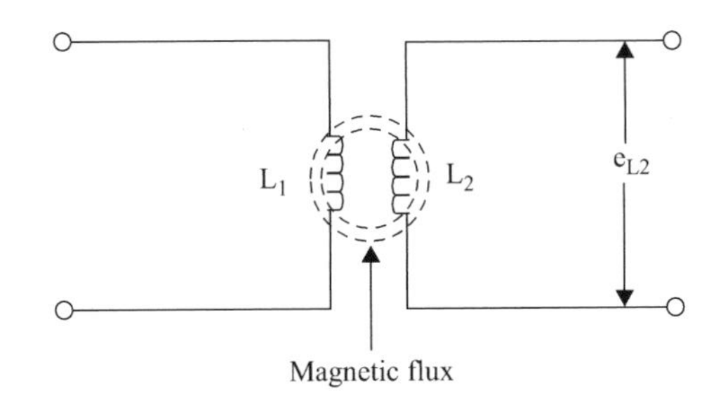

FIGURE 5.17 Mutual inductance between L_1 and L_2.

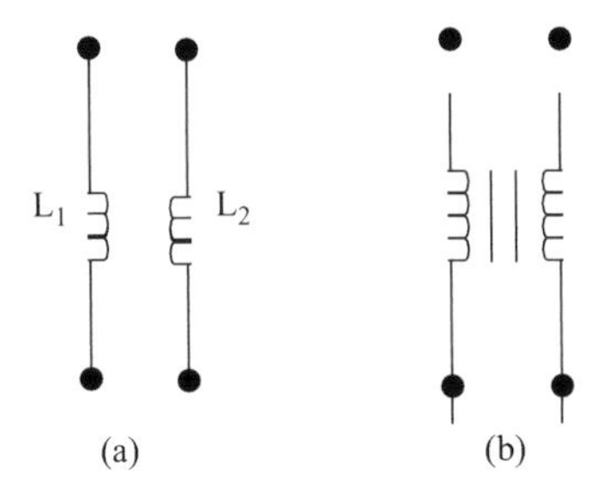

FIGURE 5.18 (a) Schematic symbol for two coils (air core) with mutual inductance; (b) two coils (iron core) with mutual inductance.

The factors affecting the mutual inductance of two adjacent coils are dependent upon

- physical dimensions of the two coils
- number of turns in each coil
- distance between the two coils
- relative positions of the axes of the two coils
- the permeability of the cores

Important Point: The amount of mutual inductance depends on the relative position of the two coils. If the coils are separated a considerable distance, the amount of flux common to both coils is small and the mutual inductance is low. Conversely, if the coils are close together so that nearly all the flow of one coil links the turns of the other, mutual inductance is high. The mutual inductance can be increased greatly by mounting the coils on a common iron core.

Calculation of Total Inductance

Note: In the study of advanced electrical theory, it is necessary to know the effect of mutual inductance in solving for total inductance in both series and parallel circuits. However, for our purposes, in this manual we do not attempt to make these calculations. Instead, we discuss the basic total inductance calculations that the electric vehicle-on-wheels maintenance technician should be familiar with.

If inductors in series are located far enough apart, or well shielded to make the effects of mutual inductance negligible, the total inductance is calculated in the same manner as for resistances in series; we merely add them:

$$L_t = L_1 + L_2 + L_3 \cdots (\text{etc.}) \tag{5.13}$$

Example 5.1

Problem:

If a series circuit contains three inductors whose values are 40 μh, 50 μh, and 20 μh, what is the total inductance?

Solution:

$$\begin{aligned} L_t &= 40\,\mu\text{h} + 50\,\mu\text{h} + 20\,\mu\text{h} \\ &= 110\,\mu\text{h} \end{aligned}$$

In a parallel circuit containing inductors (without mutual inductance), the total inductance is calculated in the same manner as for resistances in parallel:

$$\frac{1}{L_t} = \frac{1}{L_1} + \frac{1}{L_2} + \frac{1}{L_3} + \cdots (\text{etc.}) \tag{5.14}$$

Example 5.2

Problem:

A circuit contains three totally shielded inductors in parallel. The values of the three inductances are: 4 mh, 5 mh, and 10 mh. What is the total inductance?

Solution:

$$\frac{1}{L_t} = \frac{1}{4} + \frac{1}{5} + \frac{1}{10}$$

$$= 0.25 + 0.2 + 0.1$$

$$= 0.55$$

$$L_t = \frac{1}{0.55}$$

$$= 1.8 \text{ mh}$$

Capacitance

No matter how complex the electrical circuit, it is composed of no more than three basic electrical properties: resistance, inductance, and capacitance. Accordingly, gaining a thorough understanding of these three basic properties is a necessary step toward the understanding of electrical equipment. Because resistance and inductance have been covered, the last of the basic three, capacitance, is covered in this section.

Earlier, we learned that inductance opposes any change in current. *Capacitance* is the property of an electric circuit that opposes any change of *voltage* in a circuit. That is, if applied voltage is increased, capacitance opposes the change and delays the voltage increase across the circuit. If applied voltage is decreased, capacitance tends to maintain the higher original voltage across the circuit, thus delaying the decrease.

Capacitance is also defined as that property of a circuit that enables energy to be stored in an electric field. Natural capacitance exists in many electric circuits. However, in this manual, we are concerned only with the capacitance that is designed into the circuit by means of devices called **capacitors**.

Key Point: The most noticeable effect of capacitance in a circuit is that voltage can neither increase nor decrease rapidly in a capacitive circuit as it can in a circuit that does not include capacitance.

The Capacitor

A *capacitor*, or condenser, is a manufactured electrical device that consists of two conducting plates of metal separated by an insulating material called a *dielectric* (see Figure 5.19). (Note: the prefix "di—" means "through" or "across.")

The schematic symbol for a capacitor is shown in Figure 5.20.

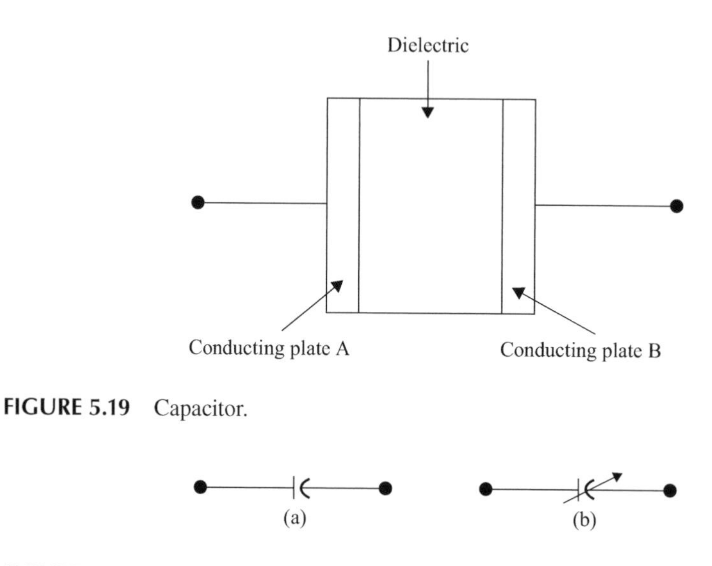

FIGURE 5.19 Capacitor.

FIGURE 5.20 (a) Schematic for a fixed capacitor; (b) Variable capacitor.

When a capacitor is connected to a voltage source, there is a short current pulse. A capacitor stores this electric charge in the dielectric (it can be charged and discharged, as we shall see later). To form a capacitor of any appreciable value, however, the area of the metal pieces must be quite large and the thickness of the dielectric must be quite small.

Key Point: A capacitor is essentially a device that stores electrical energy.

The capacitor is used in a number of ways in electrical circuits. It may block DC portions of a circuit since it is effectively a barrier to direct current (but not to AC current). It may be part of a tuned circuit—one such application is in the tuning of a radio to a particular station. It may be used to filter AC out of a DC circuit. Most of these are advanced applications that are beyond the scope of this presentation; however, a basic understanding of capacitance is necessary to the fundamentals of AC theory.

Important Point: A capacitor does not conduct DC current. The insulation between the capacitor plates blocks the flow of electrons. We learned earlier there is a short current pulse when we first connect the capacitor to a voltage source. The capacitor quickly charges to the supply voltage, and then the current stops.

The two plates of the capacitor shown in Figure 5.21 are electrically neutral since there are as many protons (positive charge) as electrons (negative charge) on each plate. Thus the capacitor has **no charge**.

Now a battery is connected across the plates (see Figure 5.22a). When the switch is closed (see Figure 5.22b), the negative charge on plate A is attracted to the positive terminal of the battery. This movement of charges will continue until the difference in charge between plates A and B is equal to the electromotive force (voltage) of the battery.

The capacitor is now **charged**. Because almost none of the charge can cross the space between plates, the capacitor will remain in this condition even if the battery is

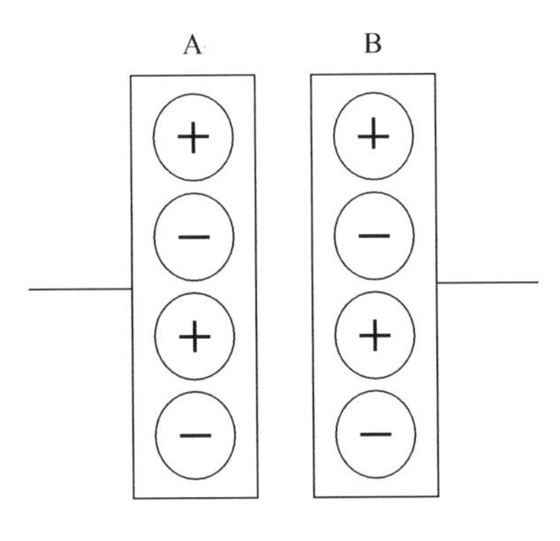

FIGURE 5.21 Two plates of a capacitor with a neutral charge.

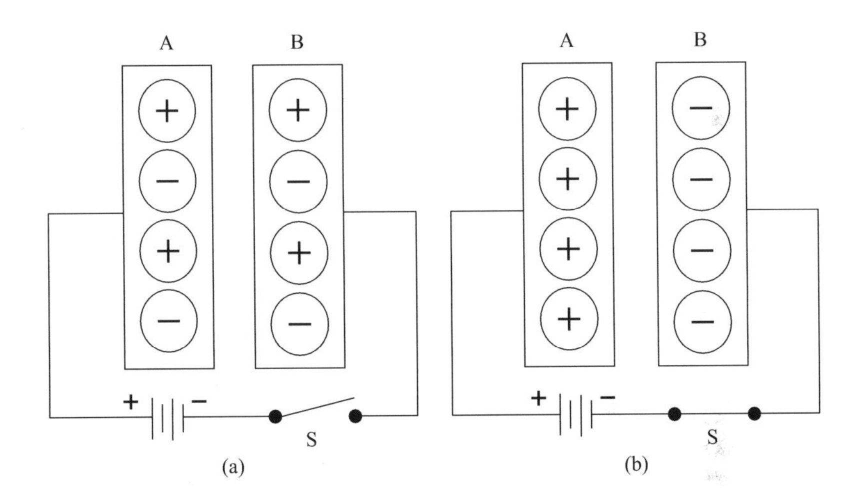

FIGURE 5.22 (a) Neutral capacitor; (b) charged capacitor.

removed (see Figure 5.23a). However, if a conductor is placed across the plates (see Figure 5.23b), the electrons find a path back to plate A and the charges on each plate are again neutralized. The capacitor is now **discharged**.

Important Point: In a capacitor, electrons cannot flow through the dielectric, because it is an insulator. Because it takes a definite quantity of electrons to charge ("fill up") a capacitor, it is said to have **capacity**. This characteristic is referred to as *capacitance.*

Dielectric Materials

Somewhat similar to the phenomenon of permeability in magnetic circuits, various materials differ in their ability to support electric flux (lines of force) or to serve as dielectric material for capacitors. Materials are rated in their ability to support electric flux in terms of a number called a *dielectric constant.* Other factors being equal,

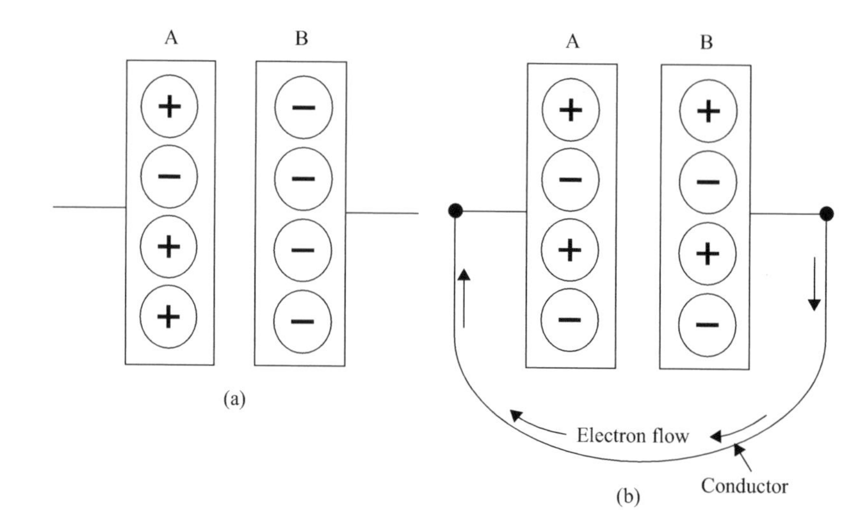

FIGURE 5.23 (a) Charged capacitor; (b) discharging a capacitor.

the higher the value of the dielectric constant, the better is the dielectric material. Dry air is the standard (the reference) by which other materials are rated.

Dielectric constants for some common materials are given in Table 5.2.

Note: From Table 5.2, it is obvious that "pure" water is the best dielectric. Keep in mind that the key word is "pure." Water capacitors are used today in some high-energy applications, in which differences in potential are measured in thousands of volts.

TABLE 5.2
Dielectric Constants

Material	Constant
Vacuum	1.0000
Air	1.0006
Paraffin paper	3.5
Glass	5–10
Quartz	3.8
Mica	3–6
Rubber	2.5–35
Wood	2.5–8
Porcelain	5.1–5.9
Glycerine (15°C)	56
Petroleum	2
Pure water	81

Unit of Capacitance

Capacitance is equal to the amount of charge that can be stored in a capacitor divided by the voltage applied across the plates:

$$C = \frac{Q}{E} \tag{5.15}$$

where
C = capacitance, F (farads)
Q = amount of charge, C (coulombs)
E = voltage, V

Example 5.3

Problem:

What is the capacitance of two metal plates separated by 1 cm of air, if 0.002 coulomb of charge is stored when a potential of 300 V is applied to the capacitor?

Solution:

Given: $Q = 0.001$ coulomb
$F = 200\,V$

$$C = \frac{Q}{E}$$

Converting to power of ten

$$C = \frac{10 \times 10^{-4}}{2 \times 10^{2}}$$

$$C = 5 \times 10^{-6}$$

$$C = 0.000005 \text{ farads}$$

Note: Although the capacitance value obtained in Example 5.3 appears small, many electronic circuits require capacitors of much smaller value. Consequently the farad is a cumbersome unit, far too large for most applications. The **microfarad,** which is one millionth of a farad (1×10^{-6} farad), is a more convenient unit. The symbols used to designate microfarad are μF.

Equation (5.15) can be rewritten as follows:

$$Q = CE \tag{5.16}$$

$$E = \frac{Q}{C} \tag{5.17}$$

Important Point: From Equation (5.16), do not deduce the mistaken idea that capacitance is dependent upon charge and voltage. Capacitance is determined entirely by physical factors, which are covered later.

The symbol used to designate a capacitor is (C). The unit of capacitance is the farad (F). The farad is that capacitance that will store one coulomb of charge in the dielectric when the voltage applied across the capacitor terminals is 1 V.

Factors Affecting the Value of Capacitance

The capacitance of a capacitor depends on three main factors: plate surface area; distance between plates; and dielectric constant of the insulating material.

- **Plate surface area**—capacitance varies directly with place surface area. We can double the capacitance value by doubling the capacitor's plate surface area. Figure 5.24 shows a capacitor with a small surface area and another one with a large surface area.

  Adding more capacitor plates can increase the plate surface area. Figure 5.25 shows alternate plates connecting to opposite capacitor terminals.

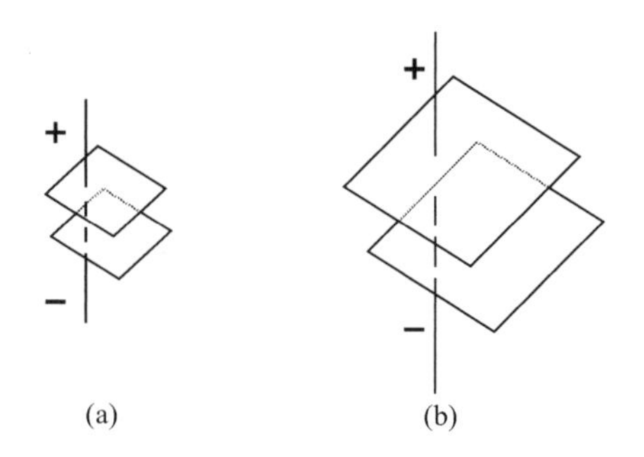

FIGURE 5.24 (a) Small plates, small capacitance; (b) larger plates, higher capacitance.

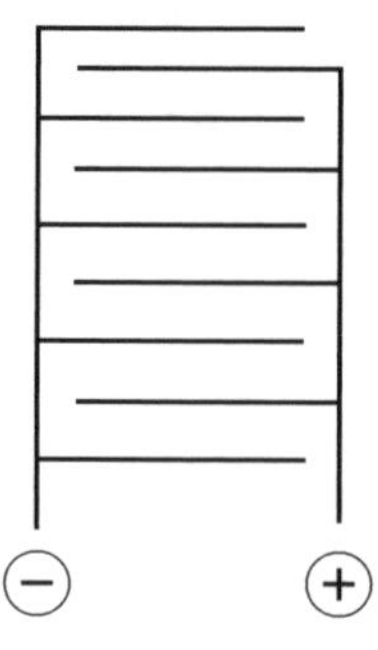

FIGURE 5.25 Several sets of plates connected to produce a capacitor with more surface area.

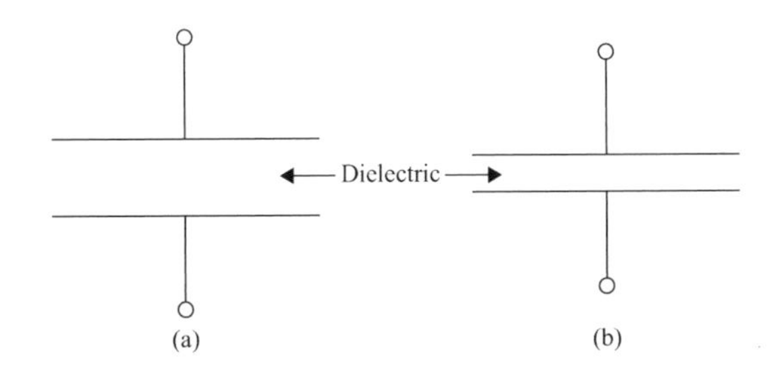

FIGURE 5.26 (a) Wide plate spacing, small capacitance; (b) narrow plate spacing, larger capacitance.

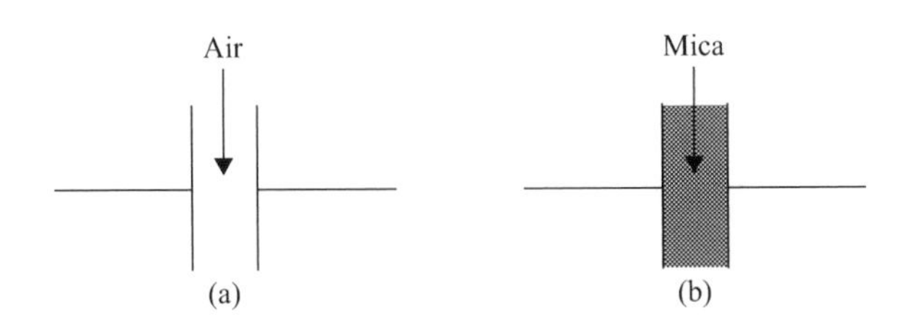

FIGURE 5.27 (a) Low capacitance; (b) higher capacitance.

- **Distance between plates**—capacitance varies inversely with the distance between plate surfaces. The capacitance increases when the plates are closer together. Figure 5.26 shows capacitors with the same plate surface area, but with different spacing.
- **Dielectric constant of the insulating material**—an insulating material with a higher dielectric constant produces a higher capacitance rating. Figure 5.27 shows two capacitors. Both have the same plate surface area and spacing. Air is the dielectric in the first capacitor and mica is the dielectric in the second one. Mica's dielectric constant is 5.4 times greater than air's dielectric constant. The mica capacitor has 5.4 times more capacitance than the air-dielectric capacitor.

Voltage Rating of Capacitors

There is a limit to the voltage that may be applied across any capacitor. If too large a voltage is applied, it will overcome the resistance of the dielectric and a current will be forced through it from one plate to the other, sometimes burning a hole in the dielectric. In this event, a short circuit exists and the capacitor must be discarded. The maximum voltage that may be applied to a capacitor is known as the **working voltage** and must never be exceeded.

The working voltage of a capacitor depends on (1) the type of material used as the dielectric and (2) the thickness of the dielectric. As a margin of safety, the capacitor should be selected so that its working voltage is at least 50% greater than the highest voltage to be applied to it. For example, if a capacitor is expected to have a maximum of 200 V applied to it, its working voltage should be at least 300 V.

Charge and Discharge of an RC Series Circuit

According to Ohm's Law, the voltage across a resistance is equal to the current through it times the value of the resistance. This means that a voltage will be developed across a resistance **only when current flows through it**.

As previously stated, a capacitor is capable of storing or holding a charge of electrons. When uncharged, both plates contain the same number of free electrons. When charged, one plate contains more free electrons than the other. The difference in the number of electrons is a measure of the charge on the capacitor. The accumulation of this charge builds up a voltage across the terminals of the capacitor, and the charge continues to increase until this voltage equals the applied voltage. The greater the voltage, the greater the charge on the capacitor. Unless a discharge path is provided, a capacitor keeps its charge indefinitely. Any practical capacitor, however, has some leakage through the dielectric so that the voltage will gradually leak off.

A voltage divider containing resistance and capacitance may be connected in a circuit by means of a switch, as shown in Figure 5.28. Such a series arrangement is called an **RC series circuit**.

If S1 is closed, electrons flow counterclockwise around the circuit containing the battery, capacitor, and resistor. This flow of electrons ceases when C is charged to the battery voltage. At the instant current begins to flow, there is no voltage on the capacitor and the drop across R is equal to the battery voltage. The initial charging current, I, is therefore equal to E_S/R.

The current flowing in the circuit soon charges the capacitor. Because the voltage on the capacitor is proportional to its charge, a voltage, e_C, will appear across the capacitor. This voltage opposes the battery voltage—that is, these two voltages buck each other. As a result, the voltage e_r across the resistor is $E_S - e_C$, and this is equal to the voltage drop ($i_C R$) across the resistor. Because E_S is fixed, i_C decreases as e_C increases.

The charging process continues until the capacitor is fully charged and the voltage across it is equal to the battery voltage. At this instant, the voltage across is zero and no current flows through it.

If S2 is closed (S1 opened) in Figure 5.28, a discharge current, i_d, will discharge the capacitor. Because i_t is opposite in direction to i_C, the voltage across the resistor will have a polarity opposite to the polarity during the charging time. However, this voltage will have the same magnitude and will vary in the same manner. During discharge the voltage across the capacitor is equal and opposite to the drop across

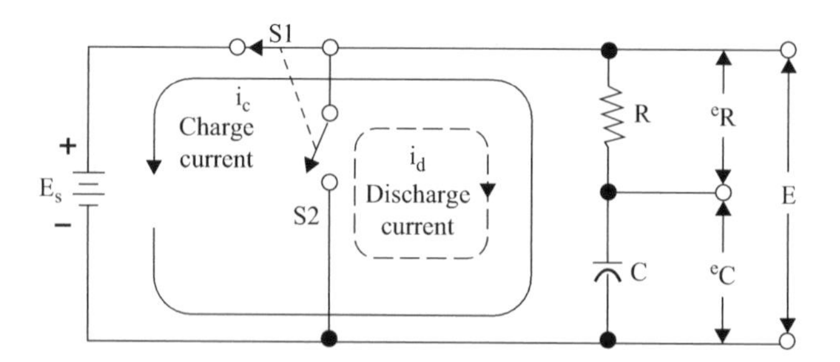

FIGURE 5.28 Charge and discharge of an RC series circuit.

the resistor. The voltage drops rapidly from its initial value and then approaches zero slowly.

The actual time it takes to charge or discharge is important in advanced electricity and electronics. Because the charge or discharge time depends on the values of resistance and capacitance, an RC circuit can be designed for the proper timing of certain electrical events. RC time constant is covered in the next section.

RC Time Constant

The time required to charge a capacitor to 63% of maximum voltage or to discharge it to 37% of its final voltage is known as the *time constant* of the current. An RC circuit is shown in Figure 5.29.

The time constant T for an RC circuit is

$$T = RC \tag{5.18}$$

The time constant of an RC circuit is usually very short because the capacitance of a circuit may be only a few microfarads or even picofarads.

Key Point: An RC time constant expresses the charge and discharge times for a capacitor.

Capacitors in Series and Parallel

Like resistors or inductors, capacitors may be connected in series, in parallel, or in a series-parallel combination. Unlike resistors or inductors, however, total capacitance in series, in parallel, or in a series-parallel combination is found in a different manner. Simply put, the rules are not the same for the calculation of total capacitance. This difference is explained as follows:

Parallel capacitance is calculated like series resistance, and series capacitance is calculated like parallel resistance. For example:

When capacitors are connected in **series** (see Figure 5.30), the total capacitance C_T is

$$\text{Series}: \frac{1}{C_T} = \frac{1}{C_1} + \frac{1}{C_2} + \frac{1}{C_3} + \cdots \frac{1}{C_n} \tag{5.19}$$

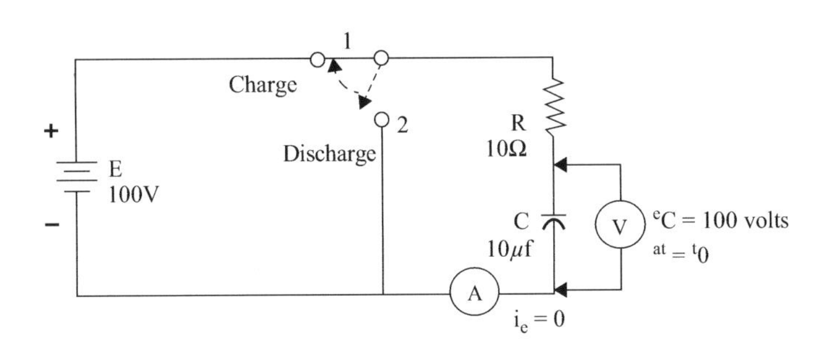

FIGURE 5.29 RC circuit.

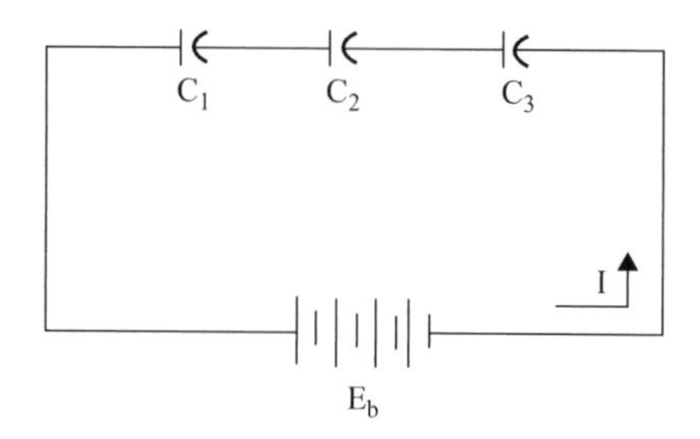

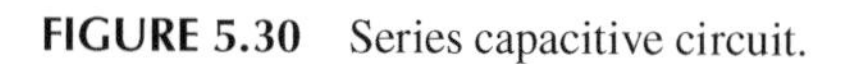
FIGURE 5.30 Series capacitive circuit.

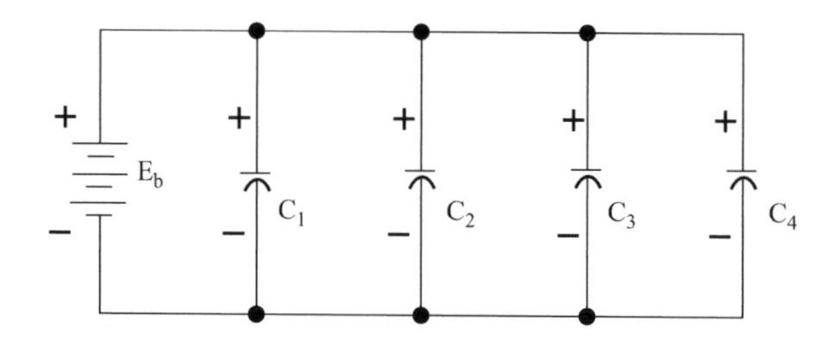

FIGURE 5.31 Parallel capacitive circuit.

Example 5.4

Problem:

Find the total capacitance of a 3-μF, a 5-μF, and a 15-μF capacitor in series.

Solution:

Write Equation (5.19) for three capacitors in series.

$$\frac{1}{C_T} = \frac{1}{C_1} + \frac{1}{C_2} + \frac{1}{C_3}$$

$$= \frac{1}{3} + \frac{1}{5} + \frac{1}{15} = \frac{9}{15} = \frac{3}{5} = \frac{5}{3} = 1.7\,\text{F}$$

When capacitors are connected in **parallel** (see Figure 5.31), the total capacitance C_T is the sum of the individual capacitances.

$$\text{Parallel}: C_T = C_1 + C_2 + C_3 + \cdots + C_n \tag{5.20}$$

Example 5.5

Problem:

Determine the total capacitance in a parallel capacitive circuit:

Given: $C_1 = 2\ \mu F$
$C_2 = 3\ \mu F$
$C_3 = 0.25\ \mu F$

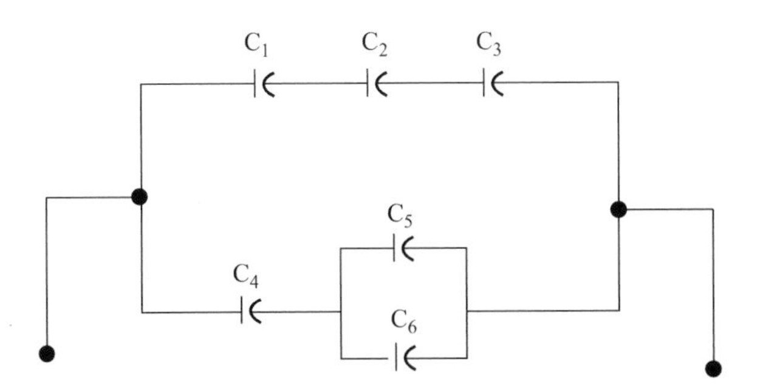

FIGURE 5.32 Series-parallel capacitance configuration.

Solution:

Write Equation (5.20) for three capacitors in parallel:

$$\begin{aligned} C_T &= C_1 + C_2 + C_3 \\ &= 2 + 3 + 0.25 \\ &= 5.25\ \mu F \end{aligned}$$

Capacitors can be connected in a combination of **series and parallel** (see Figure 5.32).

Types of Capacitors

Capacitors used for commercial applications are divided into two major groups—fixed and variable—and are named according to their dielectric. Most common are air, mica, paper, and ceramic capacitors, plus the electrolytic type. These types are compared in Table 5.3.

The fixed capacitor has a set value of capacitances that is determined by its construction. The construction of the variable capacitor allows a range of capacitances. Within this range, the desired value of capacitance is obtained by some mechanical

TABLE 5.3
Comparison of Capacitor Types

Dielectric	Construction	Capacitance Range
Air	Meshed plates	10–400 pF
Mica	Stacked plates	10–5,000 pF
Paper	Rolled foil	0.001–1 μF
Ceramic	Tubular	0.5–1,600 pF
	Disk	0.002–0.1 μF
Electrolytic	Aluminum	5–1,000 μF
	Tantalum	0.01–300 μF

means, such as by turning a shaft (as in turning a radio tuner knob, for example) or adjusting a screw to adjust the distance between the plates.

The electrolytic capacitor consists of two metal plates separated by an electrolyte. The electrolyte, either paste or liquid, is in contact with the negative terminal, and this combination forms the negative electrode. The dielectric is a very thin film of oxide deposited on the positive electrode, which is aluminum sheet. Electrolytic capacitors are polarity sensitive (i.e., they must be connected in a circuit according to their polarity markings) and are used where a large amount of capacitance is required.

Ultracapacitors

NREL (2000) points out that like batteries, ultracapacitors, also known as supercapacitors, pseudocapacitor, electric double-layer capacitors, or electrochemical double-layer capacitors (EDLCs), are energy storage devices. To meet the power, energy, and voltage requirements for a wide range of applications, they use electrolytes and configure various-sized cells into modules. As storage devices, however, they differ from batteries in that ultracapacitors (which are true capacitors in that energy is stored via charge separation at the electrode-electrolyte interface) store energy electrostatically, whereas batteries store energy chemically. Not only do ultracapacitors provide quick bursts of energy, they also are an improvement in about two to three orders of magnitude in capacitance (as compared to an average capacitor), but with a lower working voltage. Moreover, they can withstand hundreds or thousands of charge/discharge cycles without degrading. As an alternative energy source, ultracapacitors have proven themselves as reliable energy storage components used to power a variety of electronic and portable devices such as AM/FM radios, flashlights, cellphones, and emergency kits. As ultracapacitor technology matures they are being developed to function as batteries; for example, the vehicle industry is deploying ultracapacitors as a replacement for chemical batteries.

In the automotive industry, they have been integrated into electric (EV) and hybrid electric vehicles (HEV) to help alleviate stress and extend the life of the batteries.

Ultracapacitor Operation

An ultracapacitor polarizes an electrolytic solution to store energy electrostatically. Though it is an electrochemical device, no chemical reactions are involved in its energy storage mechanism. This mechanism is highly reversible and allows the ultracapacitor to be charged and discharged hundreds or thousands of times.

An ultracapacitor can be viewed as two nonreactive porous plates, or collectors, suspended within an electrolyte, with a voltage potential applied across the collectors. In an individual ultracapacitor cell, the applied potential on the positive electrode attracts the negative ions in the electrolyte, while the potential on the negative electrode attracts the positive ions. A dielectric separator between the two electrodes prevents the charge from moving between the two electrodes.

Once the ultracapacitor is charged and energy stored, a load can use this energy. The amount of energy stored is very large compared to a standard capacitor because of the enormous surface area created by the porous carbon electrodes and the small separation (10 angstroms) created by the dielectric separate. However, it stores a much smaller amount of energy than does a battery. Because the rates of charge and discharge are determined solely by its physical properties, the ultracapacitor can release energy much faster (with more power) than a battery that relies on slow chemical reactions.

INDUCTIVE AND CAPACITIVE REACTANCE

Earlier, we learned that the inductance of a circuit acts to oppose any change of current flow in that circuit and that capacitance acts to oppose any change of voltage. In DC circuits these **reactions** are not important, because they are momentary and occur only when a circuit is first closed or opened. In AC circuits these effects become very important because the direction of current flow is reversed many times each second; and the opposition presented by inductance and capacitance is, for practical purposes, constant.

In purely resistive circuits, either DC or AC, the term for opposition to current flow is resistance. When the effects of capacitance or inductance are present, as they often are in AC circuits, the opposition to current flow is called *reactance.* The total opposition to current flow in circuits that have both resistance and reactance is called *impedance.*

In this section, we cover the calculation of inductive and capacitive reactance and impedance; the phase relationships of resistance, inductive, and capacitive circuits; and power in reactive circuits.

INDUCTIVE REACTANCE

In order to gain understanding of the reactance of a typical coil, we need to review exactly what occurs when AC voltage is impressed across the coil.

1. The AC voltage produces an alternating current.
2. When a current flows in a wire, lines of force are produced around the wire.
3. Large currents produce many lines of force; small currents produce only a few lines of force.
4. As the current changes, the number of lines of force will change. The field of force will seem to expand and contract as the current increases and decreases as shown in Figure 5.33.
5. As the field expands and contracts, the lines of force must cut across the wires that form the turns of the coil.
6. These cuttings induce an emf in the coil.
7. This emf acts in the direction so as to oppose the original voltage and is called a *counter*, or back, *emf.*
8. The effect of this counter emf is to reduce the original voltage impressed on the coil. The net effect will be to reduce the current below that which would flow if there were no cuttings or counter emf.
9. In this sense, the counter emf is acting as a resistance in reducing the current.

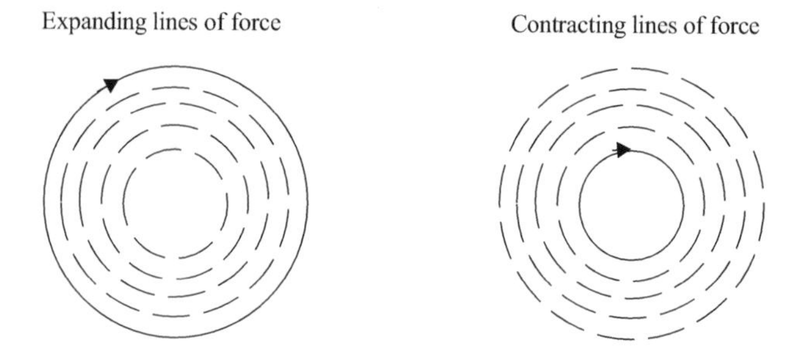

FIGURE 5.33 An AC current producing a moving (expanding and collapsing) field. In a coil, this moving field cuts the wires of the coil.

10. Although it would be more convenient to consider the current-reducing effect of a counter emf as a number of ohms of effective resistance, we don't do this. Instead, since a counter emf is not actually a resistance but merely **acts** as a resistance, we use the term *reactance* to describe this effect.

Important Point: The **reactance** of a coil is the number of ohms of resistance that the coil **seems** to offer as a result of a counter emf induced in it. Its symbol is X to differentiate it from the DC resistance R.

The inductive reactance of a coil depends primarily on (1) the coil's inductance and (2) the frequency of the current flowing through the coil. The value of the reactance of a coil is therefore proportional to its inductance and the frequency of the AC circuit in which it is used.

The formula for inductive reactance is

$$X_L = 2\pi fL \tag{5.21}$$

Since $2\pi = 2(3.14) = 6.28$, Equation (5.21) becomes

$$X_L = 6.28fL$$

where

X_L = inductive reactance, Ω
f = frequency, Hz
L = inductance, H

If any two quantities are known in Equation (5.21), the third can be found.

$$L = \frac{X_L}{6.28f} \tag{5.22}$$

$$f = \frac{X_L}{6.28L} \tag{5.23}$$

Example 5.6

Problem:

The frequency of a circuit is 60 Hz and the inductance is 20 mh. What is X_L?

Solution:

$$X_L = 2\pi fL$$
$$= 6.28 \times 60 \times 0.02$$
$$= 7.5\,\Omega$$

Example 5.7

Problem:

A 30-mh coil is in a circuit operating at a frequency of 1,400 kHz. Find its inductive reactance.

Solution:

Given: L = 30 mh
f = 1,400 kHz
Find XL =?

Step 1: Change units of measurement.

$$30\text{ mh} = 30 \times 10^{-3}\text{ h}$$
$$1{,}400\text{ kHz} = 1{,}400 \times 10^{3}$$

Step 2: Find the inductive reactance.

$$X_L = 6.28fL$$
$$X_L = 6.28 \times 1{,}400 \times 10^{3} \times 30 \times 10^{-3}$$
$$X_L = 263{,}760\,\Omega$$

Example 5.8

Problem:

Given: L = 400 μh
f = 1500 Hz
Find XL =?

Solution:

$$X_L = 2\pi fL$$
$$= 6.28 \times 1{,}500 \times 0.0004$$
$$= 3.78\,\Omega$$

Key Point: If frequency or inductance varies, inductive reactance must also vary. A coil's inductance does not vary appreciably after the coil is manufactured unless it is designed as a variable inductor. Thus, frequency is generally the only variable factor affecting the inductive reactance of a coil. The coil's inductive reactance will vary directly with the applied frequency.

Capacitive Reactance

Previously, we learned that as a capacitor is charged, electrons are drawn from one plate and deposited on the other. As more and more electrons accumulate on the second plate, they begin to act as an opposing voltage, which attempts to stop the flow of electrons just as a resistor would do. This opposing effect is called the *reactance* of the capacitor and is measured in ohms. The basic symbol for reactance is X, and the subscript defines the type of reactance. In the symbol for inductive reactance, X_L, the subscript L refers to inductance. Following the same pattern, the symbol for capacitive reactance is X_C.

Key Point: **Capacitive reactance, X_C,** is the opposition to the flow of AC current due to the capacitance in the circuit.

The factors affecting capacitive reactance, X_C, are:

- the size of the capacitor
- frequency

The larger the capacitor, the greater the number of electrons that may be accumulated on its plates. However, because the plate area is large, the electrons do not accumulate in one spot but spread out over the entire area of the plate and do not impede the flow of new electrons onto the plate. Therefore, a large capacitor offers a small reactance. If the capacitance were small, as in a capacitor with a small plate area, the electrons could not spread out and would attempt to stop the flow of electrons coming to the plate. Therefore, a small capacitor offers a large reactance. The reactance is therefore **inversely** proportional to the capacitance.

If an AC voltage is impressed across the capacitor, electrons are accumulated first on one plate and then on the other. If the frequency of the changes in polarity is low, the time available to accumulate electrons will be large. This means that a large number of electrons will be able to accumulate, which will result in a large opposing effect, or a large reactance. If the frequency is high, the time available to accumulate electrons will be small. This means that there will be only a few electrons on the plates, which will result in a small opposing effect, or a small reactance. The reactance is, therefore, **inversely** proportional to the frequency.

The formula for capacitive reactance is

$$X_C = \frac{1}{2\pi fC} \tag{5.24}$$

with C measured in farads.

Example 5.9

Problem:

What is the capacitive reactance of a circuit operating at a frequency of 60 Hz, if the total capacitance is 130 μF?

Solution:

$$X_C = \frac{1}{2\pi fC}$$

$$= \frac{1}{6.28 \times 60 \times 0.00013}$$

$$= 20.4\ \Omega$$

Phase Relationship of R, L, and C Circuits

Unlike a purely resistive circuit (where current rises and falls with the voltage; that is, it neither leads nor lags and current and voltage are in phase), current and voltage are not in phase in inductive and capacitive circuits. This is the case, of course, because occurrences are not quite instantaneous in circuits that have either inductive or capacitive components.

In the case of an inductor, voltage is first applied to the circuit, then the magnetic field begins to expand, and self-induction causes a counter current to flow in the circuit, opposing the original circuit current. Voltage **leads** current by 90° (see Figure 5.34).

When a circuit includes a capacitor, a charge current begins to flow and then a difference in potential appears between the plates of the capacitor. Current **leads** voltage by 90° (see Figure 5.35).

Key Point: In an inductive circuit, voltage leads current by 90°; and in a capacitive circuit, current leads voltage by 90°.

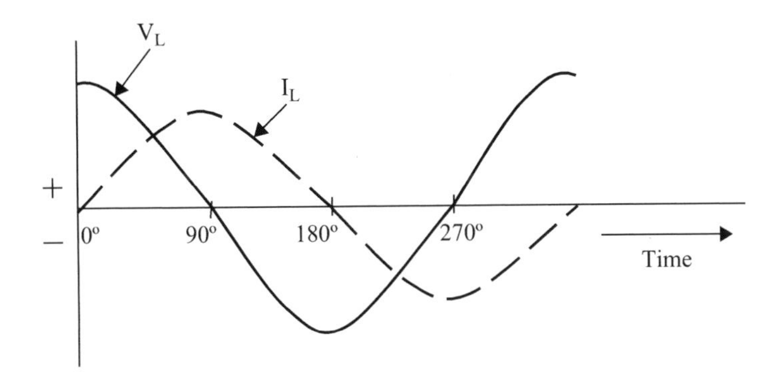

FIGURE 5.34 Inductive circuit—voltage leads current by 90°.

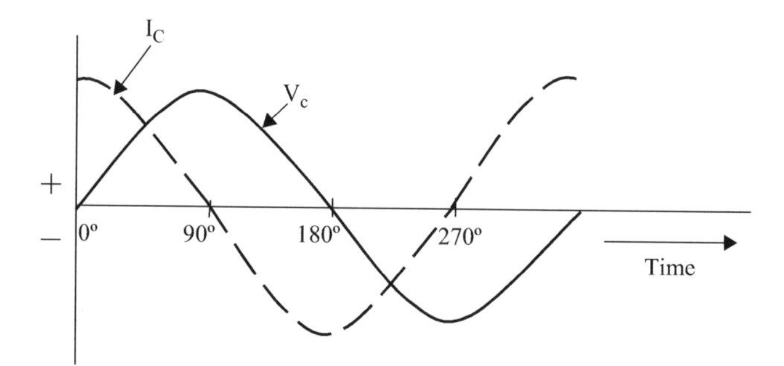

FIGURE 5.35 Capacitive circuit—current leads voltage by 90°.

IMPEDANCE

Impedance is the total opposition to the flow of alternating current in a circuit that contains resistance and reactance. In the case of pure inductance, inductive reactance X_L is the total opposition to the flow of current through it. In the case of pure resistance, R represents the total opposition. The combined opposition of R and X_L in series or in parallel to current flow is called **impedance**. The symbol for impedance is Z.

The impedance of resistance in series with inductance is

$$Z = \sqrt{R^2 + X_L^{\ 2}} \tag{5.25}$$

where

Z = Impedance, Ω
R = Resistance, Ω
X_L = Inductive Reactance, Ω

The impedance of resistance in series with capacitance is:

$$Z = \sqrt{R^2 + X_C^{\ 2}} \tag{5.26}$$

where

Z = Impedance, Ω
R = Resistance, Ω
X_C = Inductive Capacitance, Ω

When the impedance of a circuit includes R, X_L, and X_C, both resistance and net reactance must be considered. The equation for impedance, including both X_L and X_C, is:

$$Z = \sqrt{R^2 + (X_L - X_C)^2} \tag{5.27}$$

Power in Reactive Circuits

The power in a DC circuit is equal to the product of volts and amps, but in an AC circuit this is true only when the load is resistive and has no reactance.

In a circuit possessing inductance only, the true power is zero. The current lags the applied voltage by 90°. The true power in a capacitive circuit is also zero. The *true power* is the average power actually consumed by the circuit, the average being taken over one complete cycle of alternating current. The *apparent power* is the product of the rms volts and rms amps.

The ratio of true power to apparent power in an AC circuit is called the *power factor*. It may be expressed as a percent or as a decimal.

To this point we have pointed out how a combination of inductance and resistance and then capacitance and resistance behaves in an AC circuit. We saw how the RL and RC combination affects the current, voltages, power, and power factor of a circuit. We considered these fundamental properties as isolated phenomena. The following phase relationships were seen to be true:

1. The voltage drop across a resistor is **in phase** with the current through it.
2. The voltage drop across an inductor **leads** the current through it by 90°.
3. The voltage drop across a capacitor **lags** the current through it by 90°.
4. The voltage drops across inductors and capacitors are **180° out of phase**.

Solving AC problems is complicated by the fact that current varies with time as the AC output of an alternator goes through a complete cycle. This is the case because the various voltage drops in the circuit vary in phase—they are not at their maximum or minimum values at the same time.

AC circuits frequently include all three circuit elements: resistance, inductance, and capacitance. In this section all three of these fundamental circuit parameters are combined and their effect on circuit values studied.

Series RLC Circuit

Figure 5.36 shows both the sine waveforms and the vectors for purely resistive, inductive, and capacitive circuits. Only the vectors show the direction, because the magnitudes are dependent on the values chosen for a given circuit. (**Note**: We are only interested in the **effective** (root-mean square, rms) values.) If the individual resistances and reactance are known, Ohm's Law may be applied to find the voltage drops. For example, we know that $E_R = I \times R$ and $E_C = I \times X_L$. Then, according to Ohm's Law, $E_L = I \times X_L$.

In AC circuits, current varies with time; accordingly, the voltage drops across the various elements also vary with time. However, the same variation is not always present in each **at the same time** (except in purely resistive circuits) because current and voltage are not in phase.

Important Point: In a resistive circuit, the phase difference between voltage and current is zero.

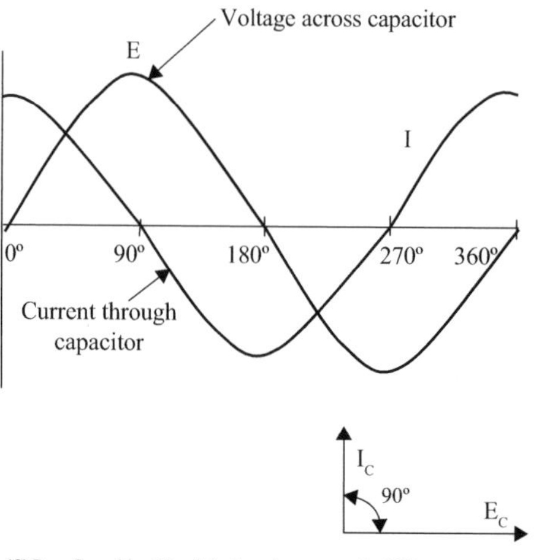

(C) Pure Capacitive Circuit (voltage lags current by 90°).

FIGURE 5.36 Sine waveforms and vectorial representation of R, L, and C circuits.

We are concerned, in practical terms, mostly with effective values of current and voltage. However, to understand basic AC theory, we need to know what occurs from instant to instant.

In Figure 5.37, note first that current is the common reference for all three element voltages, because there is only one current in a series circuit, and it is common to all

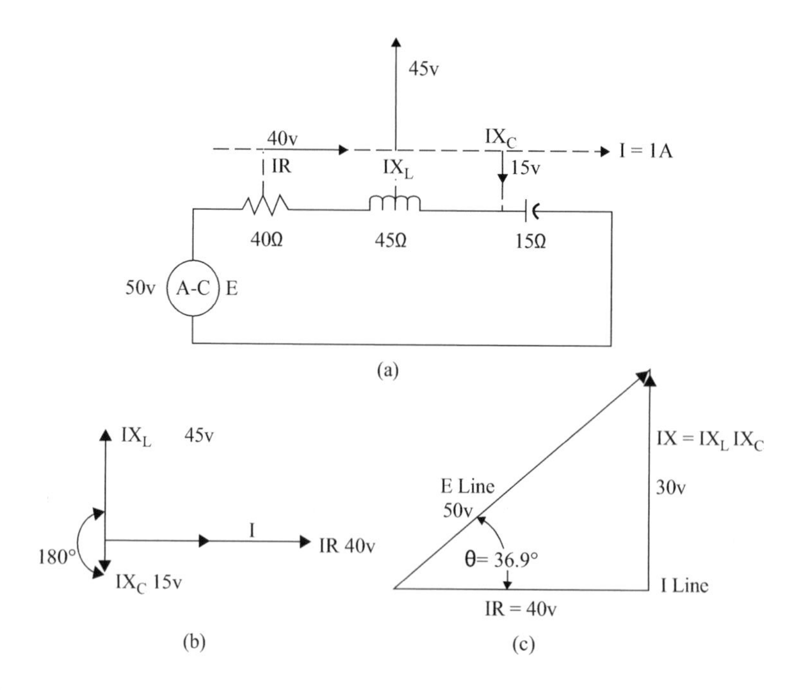

FIGURE 5.37 (a) Resistance, (b) inductance, and (c) capacitance connected in a series.

elements. The dashed line in Figure 5.37a represents the common series current. The voltage vector for each element, showing its individual relation to the common current, is drawn above each respective element. The total source voltage E is the vector sum of the individual voltages of IR, IX_L, and IX_C.

The three element voltages are arranged for summation in Figure 5.37b. Because IX_L and IX_C are each 90° away from I, they are therefore 180° from each other. Vectors in direct opposition (180° out of phase) may be subtracted directly. The total reactive voltage E_X is the difference of IX_L and IX_C. Or $E_X = IX_L - IX_C = 45 - 15 = 30\,V$.

Important Point: The voltage across a single reactive element in a series circuit can have a greater effective value than that of the applied voltage.

The final relationship of line voltage and current, as seen from the source, is shown in Figure 5.37c. Had X_C been larger than X_L, the voltage would lag, rather than lead. When X_C and X_L are of equal values, line voltage and current will be in phase.

Important Point: One of the most important characteristics of a RLC circuit is that it can be made to respond most effectively to a single given frequency. When operated in this condition, the circuit is said to be in *resonance* with or *resonant* to the operating frequency. A circuit is at resonance when the inductive reactance X_L is equal to the capacitive reactance X_C. At resonance Z equals the resistance R.

In summary, the series RLC circuit illustrates three important points:

- The current in a series RLC circuit either leads or lags the applied voltage, depending on whether X_C is greater or less than X_L.
- A capacitive voltage drop in a series circuit always subtracts directly from an inductive voltage drop.
- The voltage across a single reactive element in a series circuit can have a greater effective value than that of the applied voltage.

Parallel RLC Circuits

The **true power** of a circuit is $P = EI \cos \theta$; and for any given amount of power to be transmitted, the current, I, varies inversely with the power factor, $\cos \theta$. Thus, the addition of capacitance in parallel with inductance will, under the proper conditions, improve the power factor (make the power factor nearer to unity) of the circuit and make possible the transmission of electric power with reduced line loss and improved voltage regulation.

Figure 5.38a shows a three-branch parallel AC circuit with a resistance in one branch, inductance in the second branch, and capacitance in the third branch. The voltage is the same across each parallel branch, so $V_T = V_R = V_L = V_C$. The applied voltage V_T is used as the reference line to measure phase angle θ. The total current I_T is the vector sum of I_R, I_L, and I_C. The current in the resistance I_R is in phase with the applied voltage V_T (see Figure 5.38b). The current in the capacitor I_C leads the voltage V_T by 90°. I_L and I_C are exactly 180° out of phase and thus acting in opposite directions (see Figure 5.38b). When $I_C > I_C$, I_T lags V_T (see Figure 5.38c), so the parallel RLC circuit is considered inductive.

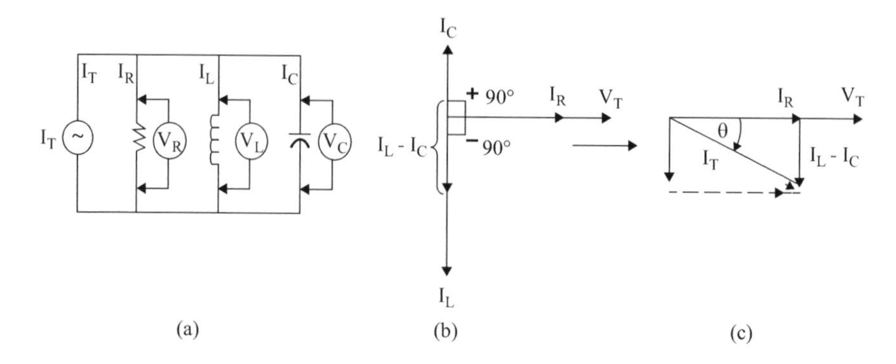

FIGURE 5.38 R, X_L, and X_C in parallel. (a) Parallel RLC circuit diagram, (b) vector diagram $I_L > I_C$, (c) current-vector triangle $I_L > I_C$.

Power in AC Circuits

In circuits that have only resistance, but no reactance, the amount of power absorbed in the circuit is easily calculated by $P = I^2R$. However, in dealing with circuits that include inductance and capacitance (or both), which is often the case in AC, the calculation of power is a more complicated process.

Earlier, we explained that power is a measure of the rate at which work is done. The "work" of a resistor is to limit current flow to the correct, safe level. In accomplishing this, the resistor dissipates heat, and we say that power is consumed or absorbed by the resistor.

Inductors and capacitors also oppose current flow, but they do so by producing current that opposes the line current. In either inductive or capacitive circuits, instantaneous values of power may be very large, but the power actually absorbed is essentially zero, since only resistance dissipates heat (absorbs power). Both inductance and capacitance return the power to the source.

Any component that has resistance, such as a resistor or the wiring of an inductor, consumes power. Such power is not returned to the source, because it is dissipated as heat. Previously, we stated that power consumed in the circuit is called **true power**, or **average power**. The two terms are interchangeable, but we use the term *average power*, because the overall value is more meaningful than the instantaneous values of power appearing in the circuit during a complete cycle.

Key Point: In terms of the dissipation of power as heat in a circuit, **apparent power** includes both power that is returned to the source and power that is dissipated as heat. **Average power** is power that is dissipated as heat.

Not all apparent power is consumed by the circuit; however, because the alternator does deliver the power, it must be considered in the design. The average power consumption may be small, but instantaneous values of voltage and current are often very large. Apparent power is an important design consideration, especially in assessing the amount of insulation necessary.

In an AC circuit that includes both reactance and resistance, some power is consumed by the load and some is returned to the source. How much of each depends on the phase angle since current normally leads or lags voltage by some angle.

Note: Recall that in a purely reactive circuit, current and voltage are 90° out of phase.

In an RLC circuit, the ratio of R/Z is the cosine of the phase angle θ. Therefore, it is easy to calculate average power in an RLC circuit:

$$P = EI\cos\theta \tag{5.28}$$

where

E = effective value of the voltage across the circuit
I = effective value of current in the circuit
θ = phase angle between voltage and current
P = average power absorbed by the circuit

Note: Recall that the equation for average power in a purely resistive circuits is P = EI. In a resistive circuit, P = EI, because cos θ is 1 and need not be considered. In most cases, the phase angle will be neither 90° nor zero, but somewhere between those extremes.

Example 5.10

Problem:

An RLC circuit has a source voltage of 500 V, line current is 2 amps, and current leads voltage by 60°. What is the average power?

Solution:

$$\text{Average Power} = 500\text{ v.} \times 2\text{ a.} \times 0.5 \text{ (Note : Cos of } 60° = 0.5)$$

$$= 500 \text{ watts}$$

Example 5.11

Problem:

An RLC circuit has a source voltage of 300 V, line current is 2 amps, and current lags voltage by 31.8°. What is the average power?

Solution:

$$\text{Average Power} = 300\text{ v.} \times 2\text{ a.} \times 0.8499$$

$$= 509.9 \text{ watts}$$

Example 5.12

Problem:

Given: E = 100 V
I = 4 amps
θ = 58.4
What is the average power?

Solution:

$$\text{Average Power} = 100 \text{ v.} \times 4 \text{ a.} \times 0.5240$$

$$= 209.6 \text{ watts}$$

REFERENCE

NREL (2000). *Ultracapacitor applications and evaluation for hybrid vehicles.* Accessed 12/12/22 @ https//www.nrel.gov/fyo9ost/45596. pdf.

6 Fundamental Physics Concepts

Education can only go so far in preparing the vehicular engineer, environmental engineer, automotive technician for on-the-job experience—the real learning occurs on-the-job. A person who wishes to become an engineer or skilled technician in any field is greatly assisted by two personal factors. First, a well-rounded, broad development of experience in many areas is required, and results in the production of the classic generalist. Although engineers cannot possibly attain great depth in all areas, they must have the desire and the aptitude to do so. They must be interested in—and well informed about—many widely differing fields of study. The necessity for this in the engineering application is readily apparent. Why? Simply because the range of problems encountered is so immense that a narrow education will not suffice; engineers must handle situations that call upon skills as widely diverse as the ability to understand psychological and sociological problems with people to the ability to perform calculations required for mechanics, structures, and vehicle dynamics. The aspiring engineer can come from just about any background, and that a narrow education does not preclude students and others from broadening themselves later; however, quite often those who are very specialized possess narrow vision, lack of far-thinking, and lack of appreciation for other disciplines, as well as the adaptability necessary to design, engineer and maintain vehicles on wheels.

A second requirement calls for educational emphasis upon quantitative and logical problem solving. The non-technician whose mathematical ability breaks down at simple algebra is not likely to acquire the necessary quantitative expertise without great effort. Along with mathematics, the engineer must have a good foundation in mechanics and structures. An education that does not include a foundation in the study of forces that act on machines, vehicular structures, mechanics, materials, dynamics and processes leaves the engineering practitioner in the same position as a thoracic surgeon with incomplete knowledge of gross anatomy—he or she simply has to feel their way to the target, leaving a lot to be desired (especially for the patient).

—Frank R. Spellman (1996)

INTRODUCTION

Though individual learning style is important in your determination of a career choice as a vehicle designer, vehicle engineer, or professional vehicle maintainer, again, we stress education as the key ingredient in the mix that produces the safe,

DOI: 10.1201/9781003332992-6

convenient, operational, vehicle-on-wheels. Along with basic and applied sciences (mathematics, natural, and behavioral sciences—which are applied to the solution of technological, biological, and behavioral problems), education and formal on-the-job practice in engineering and technology are a must for the aspiring vehicle engineer or advanced vehicle technical designer. Topics including applied mechanics, properties of materials, electrical circuits and machines, principles of engineering design, vehicle dynamics, and computer science fall into this category.

In this chapter, the focus is on applied mechanics and, in particular, forces and the resolution of forces. Why? Because many equipment failures, accidents, and resulting injuries are caused by forces of too great a magnitude for a machine, material, or structure. To design and inspect systems, devices, or products to ensure their reliability and safety, vehicle-on-wheels engineers must account for the forces that act or might act on them. Moreover, environmental engineers and others must also account for forces from objects that may act on the human body (an area of focus that is often overlooked but is critical in avoiding recalls and lawsuits). Important areas that are part of or that interface with applied mechanics are the properties of materials, electrical circuits and machines (components), and engineering design considerations. We cannot discuss all engineering aspects related to these areas in this text. Instead, our goal is to look at some fundamental concepts, and their relationships to vehicle-on-wheels engineering.

RESOLUTION OF FORCES

In engineering, we tend to focus our attention on those forces that are likely to support optimum performance, cause failure or damage to some device or system, resulting in an occurrence that is likely to produce secondary damage to other devices or systems and harm to individuals. Typically, large forces are more likely to cause failure or damage than small ones.

The vehicle-on-wheels engineer and advanced technicians must understand force. He or she must understand how a force acts on a body (vehicular and human): (1) the direction of force, (2) point of application (location) of force, (3) the area over which force acts, (4) the distribution or concentration of forces that act on bodies, and (5) how essential these elements are in evaluating the strength of materials. For example, a 40-lb force applied to the edge of a sheet of plastic and parallel to it probably will not break it. If a sledgehammer strikes the center of the sheet with the same force, the plastic will probably break. A sheet metal panel of the same size undergoing the same force will not break.

Practice tells us that different materials have different strength properties. Striking the plastic panel will probably cause it to break, whereas striking a sheet metal panel will cause a dent. The strength of a material and its ability to deform are directly related to the force applied. Important physical, mechanical, and other properties of materials are:

- crystal structure
- strength
- melting point

- density
- hardness
- brittleness
- ductility
- modulus of elasticity
- wear properties
- coefficient of expansion
- contraction
- conductivity
- shape
- exposure to environmental conditions
- exposure to chemicals
- fracture toughness

and many others. Note that all these properties can vary, depending on whether forces are crushing, corroding, cutting, pulling, rolling, or twisting.

The forces an object can encounter are often different from the forces that an object would be able to withstand. The object may be designed to withstand only minimal force before it fails (a toy doll may be designed either of very soft, pliable materials or designed to break or give way in certain places when a child falls on it, preventing injury). Other devices may be designed to withstand the greatest possible load and shock (e.g., a building constructed to withstand an earthquake).

When working with any material to go into a vehicle-on-wheels with a concern for safety, a safety factor (or factor of safety) is often introduced. Safety factor (SF) [as defined by ASSE (1988)] is the ratio allowed for in design, between the ultimate breaking strength of a member, material, structure, or equipment and the actual working stress or safe permissible load placed on it during ordinary use. Simply put, including a factor of safety—into the design of a machine, for example—makes an allowance for many unknowns (inaccurate estimates of real loads or irregularities in materials, for example) related to the materials used to make the machine, related to the machine's assembly, and related to the use of the machine. Safety factor (SF) can be determined in several ways. One of the most commonly used ways is

$$SF = \frac{\text{failure} - \text{producing load}}{\text{allowable stress}} \tag{6.1}$$

Forces on a material or object are classified by the way they act on the material. For example, if the force pulls a material apart, it is called tensile force. Forces that squeeze a material or object are called compression forces. Shear forces cut a material or object. Forces that twist a material or object are called torsional forces. Forces that cause a material or object to bend are called bending forces. A bearing force occurs when one material or object presses against or bears on another material or body.

So, what is force? Force is typically defined as any influence that tends to change the state of rest or the uniform motion in a straight line of a body. The action of an unbalanced or resultant force results in the acceleration of a body in the direction

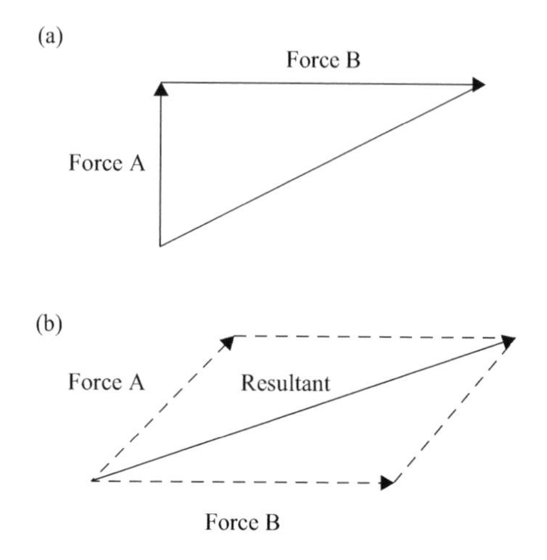

FIGURE 6.1 (a and b) Force is a vector quantity.

of action of the force, or it may (if the body is unable to move freely) result in its deformation (Hooke's Law). Force is a vector quantity, possessing both magnitude and direction (see Figure 6.1a and b); its SI unit is newton (equal to 3.6 ounces, or 0.225 lb).

According to Newton's second law of motion, the magnitude of a resultant force is equal to the rate of change of momentum of the body on which it acts; the force (F) producing acceleration (m/s^2) on a body of mass (kilograms) is therefore given by:

$$F = ma \tag{6.2}$$

With regard to environmental engineering, a key relationship between a force F and a body on which it acts is

$$F = sA \tag{6.3}$$

where

s = force or stress per unit area (e.g., pounds per square inch)
A = area (square inches, square feet, etc.) over which a force acts

Note: The stress a material can withstand is a function of the material and the type of loading.

Frequently, two or more forces act together to produce the effect of a single force, called a resultant. This resolution of forces can be accomplished in two ways: triangle and/or parallelogram law. The triangle law provides that if two concurrent forces are laid out vectorially with the beginning of the second force at the end of the first, the vector connecting the beginning and the end of the forces represents the resultant of the two forces (see Figure 6.1a). The parallelogram law provides that if two concurrent forces are laid out vectorially, with both forces pointing toward and pointing away from their point of intersection, a parallelogram represents the resultant of the force. The concurrent forces must have both direction and magnitude if their resultant is to be determined (see Figure 6.1b). After the triangle or parallelogram has been completed, and if the individual forces are known or one of the individual forces and the resultant are known, the resultant force may be simply calculated by either the trigonometric method (sines, cosines, and tangents) or by the graphic method (which involves laying out the known force, or forces, to an exact scale and exact direction in either a parallelogram or triangle, and then measuring the unknown to the same scale).

Slings

Let's take a look at a few example problems involving forces that the engineer, maintenance technician, or others might be called upon to calculate. In our examples, we use lifting slings under different conditions of loading.

Note: Slings are commonly used between cranes, derricks, and/or hoists and the load, so that the load may be lifted and moved to a desired location. For the engineer, knowledge of the properties and limitations of the slings, the type and condition of material being lifted, the weight and shape of the object being lifted, the angle of the lifting sling to the load being lifted, and the environment in which the lift is to be made are all important considerations to be evaluated—before the transfer of material can take place safely. Note that vehicles-on-wheels require assembly and maintenance including parts replacement and to accomplish these tasks slings are commonly used in one manner or another; thus, it is important to include discussion of slings in this presentation.

Example 6.1

Let us assume a load of 2,000 lb supported by a two-leg sling; the legs of the sling make an angle of 60° with the load (see Figure 6.2). What force is exerted on each leg of the sling?

Note: In solving this type of problem, you should always draw a rough diagram as shown in Figure 6.2.

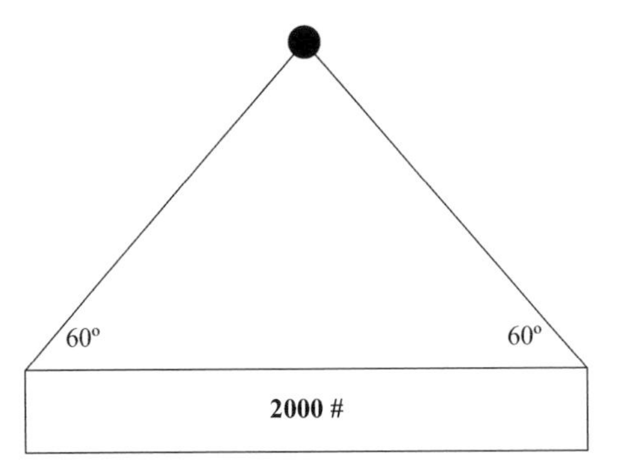

FIGURE 6.2 For Example 6.1.

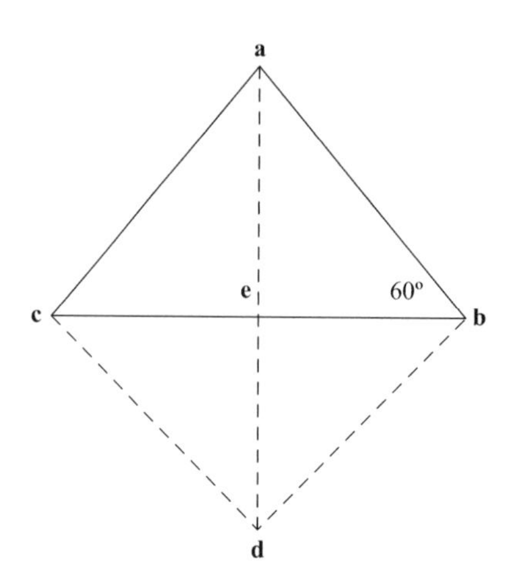

FIGURE 6.3 For Example 6.1.

A resolution of forces provides the answer. We use the trigonometric method to solve this problem but remember that it may also be solved using the graphic method. Using the trigonometric method with the parallelogram law, the problem could be solved as follows (again, make a drawing which should look like Figure 6.3):

We could consider the load (2,000 lb) as being concentrated and acting vertically, which can be indicated by a vertical line. The legs of the slings are at a 60° angle, which can be shown as **ab** and **ac**. The parallelogram can now be constructed by drawing lines parallel to **ab** and **ac**, intersecting at **d**. The point where **cb** and **ad** intersect can be indicated as **e**. The force on each leg of the sling (**ab** for example) is the resultant of two forces, one acting vertically (**ae**), the other horizontally (**be**), as shown in the force diagram. Force **ae** is equal to one-half of **ad** (the total force acting vertically, 2,000 lb, so **ae**=1,000). This value remains constant regardless of the angle **ab** makes with **bd,** because as the angle increases or decreases, **ae** also increases or decreases. But **ae** is always **ad/2**. The force **ab** can be calculated by trigonometry using the right triangle **abe:**

$$\text{Sine of an angle} = \frac{\text{opposite side}}{\text{hypotenuse}}$$

$$\text{therefore, sine 60 degrees} = \frac{\text{ae}}{\text{ab}}$$

$$\text{transposing, ab} = \frac{\text{ae}}{\text{sine 60 degrees}}$$

$$\text{substituting known value, ab} = \frac{1{,}000}{0.866} = 1{,}155$$

The total weight on each leg of the sling at a 60° angle from the load is 1,155 lb. Note that the weight is more than half the load because the load is made up of two forces—one acting vertically, the other horizontally. An important point to

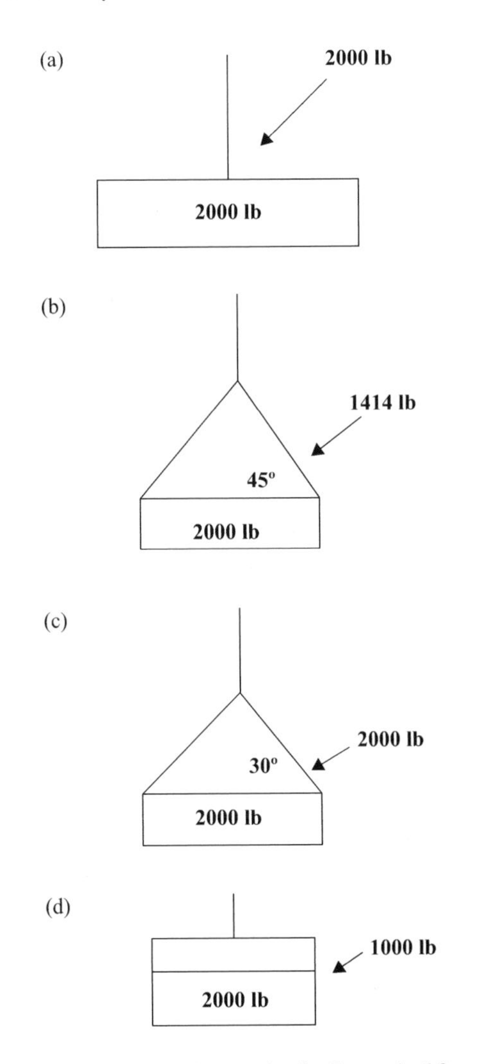

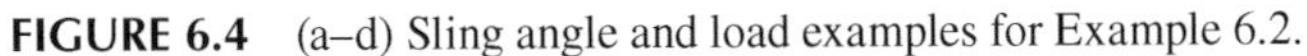
FIGURE 6.4 (a–d) Sling angle and load examples for Example 6.2.

remember is that the smaller the angle, the greater the load (force) on the sling. For example, at a 15° angle, the force on each leg of a 2,000-lb load increases to 3,864 lb.

Let's take a look at what the force would be on each leg of a 2,000-lb load at different angles (angles that are common for lifting slings; see Figure 6.4).

Now let's work a couple of example problems.

Example 6.2

Problem:

We have a 3,000-lb load to be lifted with a 2-leg sling whose legs are at a 30° angle with the load. The load (force) on each leg of the sling is:

Solution:

$$\text{Sine A} = \frac{a}{c}$$

$$\text{Sine } 30 = 0.500$$

$$a = \frac{3{,}000 \text{ lb.}}{2} = 1{,}500$$

$$c = \frac{a}{\text{Sine A}}$$

$$c = \frac{1{,}500}{0.5}$$

$$c = 3{,}000$$

Example 6.3

Problem:

Given a 2-rope sling supporting 10,000 lb, what is the load (force) on the left sling? Sling angle to load is 60°.

Solution:

$$\text{Sine A} = \frac{a}{c}$$

$$\text{Sine A} = \frac{60}{0.866}$$

$$a = \frac{10{,}000}{2}$$

$$c = \frac{a}{\text{Sine A}}$$

$$c = \frac{5{,}000}{0.866}$$

$$c = 5{,}774 \text{ lb}$$

Inclined Plane

Another common problem encountered by the engineer involves the resolution of forces occurring in material handling operations in moving a load (a vehicle chassis, for example) up an inclined plane (a ramp, in our example). The safety implications in this type of work activity should be obvious. The forces acting on an inclined plane

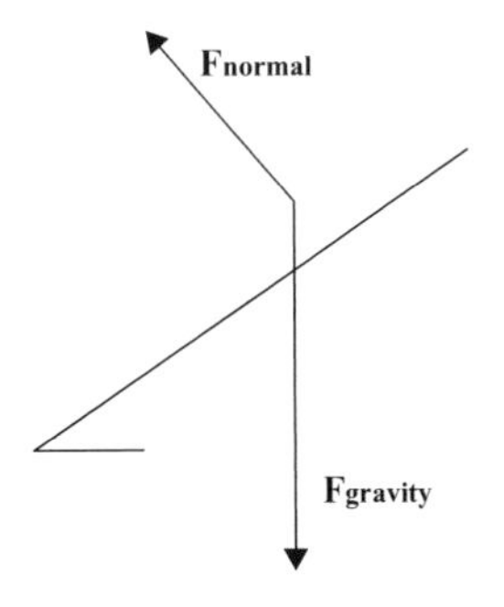

FIGURE 6.5 Forces acting on an inclined plane.

are shown in Figure 6.5. Let's take a look at a typical example of how to determine the force needed to pull a fully loaded cart up a ramp (an inclined plane).

Example 6.4

Problem:

Assume that a fully loaded cart weighing 400 lb is to be pulled up a ramp that has a 5-foot rise for each 12 feet, measured along the horizontal (again, make a rough drawing like Figure 6.6).

What force is required to pull it up the ramp?

Note: For illustrative purposes, we assume no friction. Without friction, of course, the work done in moving the cart in a horizontal direction would be zero (once the cart was started, it would move with constant velocity—the only work required is that necessary to get it started). However, a force equal to J is necessary to pull the cart up the ramp, or to maintain the car at rest (in equilibrium). As the angle (slope) of the ramp is increased, greater force is required to move it, because the load is being raised as it moves along the ramp, thus doing work (remember, this is not the case when the cart is moved along a horizontal plane without friction—in actual practice, however, friction can never be ignored and some "work" is accomplished in moving the cart).

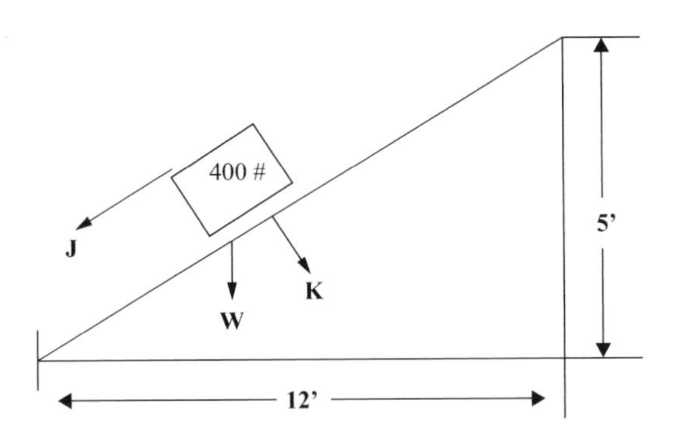

FIGURE 6.6 Inclined plane for Example 6.4.

Solution:

To determine the actual force involved, we can again use a resolution of forces. The first step is to determine the angle of the ramp. This can be calculated by the formula:

$$\text{Tangent (angle of ramp)} = \frac{\text{opposite side}}{\text{adjacent side}} = 5/12 = 0.42$$

$$\text{Then, arctan } 0.42 = 22.8 \text{ degree}$$

Now you need to draw a force parallelogram and apply the trigonometric method (see Figure 6.7). The weight of the cart W (shown as force acting vertically) can be resolved into two components, one a force J parallel to the ramp, the other a force K perpendicular to the ramp. The component K, being perpendicular to the inclined ramp (plane), does not hinder movement up the ramp. The component J represents a force that would accelerate the cart down the ramp. To pull the cart up the ramp, a force equal to or greater than J is necessary.

Applying the trigonometric method, the angle WOK is the same as the angle of the ramp.

$$\text{OJ} = \text{WK} \ \& \ \text{OW} = 400 \text{ lb}$$

$$\text{Sine of angle WOK}(22.8°) = \frac{\text{opposite side (WK)}}{\text{hypotenuse (OW)}}$$

$$\text{Transposing, WK} = \text{OW} \times \text{sine}(22.8°)$$

$$\text{WK} = 400 \times 0.388$$

$$\text{WK} = 155.2$$

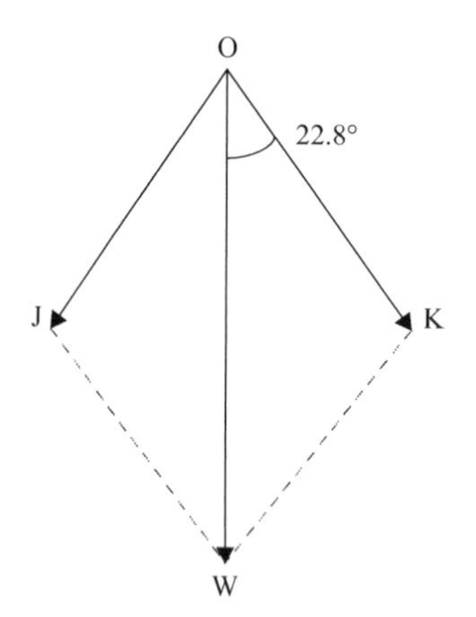

FIGURE 6.7 Force parallelogram.

Thus, a force of 155.2 lb is necessary to pull the cart up the 22.8° angle of the ramp (friction ignored). Note that the total amount of work is the same, whether the cart is lifted vertically (400 pounds × 5 feet = 2,000 foot-pounds) or pulled up the ramp (155.2 pounds × 13 feet = 2,000 foot-pounds). The advantage gained in using a ramp instead of a vertical lift is that less force is required—but through a greater distance.

PROPERTIES OF MATERIALS AND PRINCIPLES OF MECHANICS

To be able to recognize hazards and to select and implement appropriate controls, an engineer must have a good understanding of the properties of materials and principles of mechanics. In this section, we start with the properties of materials then cover the wide spectrum that is mechanics, starting with statics and ending with electrical machines. The intent is to clearly illustrate the wide scope of knowledge required in areas germane to the properties of materials, and the principles of mechanics and those topics on the periphery—all of which blend in the mix—the safety mix—the mix that helps to produce the well-rounded, knowledgeable vehicle-on-wheels engineer.

PROPERTIES OF MATERIALS

When we speak of the properties of materials, what are we referring to, and why should we concern ourselves with this topic? The best way to answer this question is to use an example where the vehicle-on-wheels engineer, working with design and safety engineers in a preliminary design conference, might typically be exposed to (should be exposed to) data, parameters, and specifications related to the properties of a particular construction material to be used in the fabrication of, for example, a large mezzanine in an assembly factory. In constructing this particular mezzanine, consideration was given to the fact that it would be used to store large, heavy equipment components. The demands placed on the finished mezzanine create the need for the mezzanine to be built using materials that can support a heavy load.

For illustration, let's say that the design engineers plan to use an aluminum alloy, type structural—No. 17ST. Before they decide to include No. 17ST and determine the required quantity needed to build the mezzanine, they are concerned with determining its mechanical properties, to ensure that it will be able to handle the intended load (they will also factor in, many times over, for safety—selecting a type of material that will handle a load much greater than expected).

Using a table on the *Mechanical Properties of Engineering Materials in Urquhart's Civil Engineering Handbook*, 4th ed., (1959) they check the following for No. 17ST (Table 6.1):

The question is: Is this information important to the engineer? No—not exactly. What is important to the engineer is: (1) that a procedure such as the one just described was actually accomplished; that is, that professional engineers actually took the time to determine the correct materials to use in constructing the mezzanine; and (2) when exposed to this type of information, to specific terms, the environmental engineer must know enough about the "language" used to know what the design engineers are

TABLE 6.1
Properties of Engineering Materials No. 17ST

Ultimate strength, psi (defined as the ultimate strength in compression for ductile materials, which is usually taken as the yield point)	Tension: 58,000 psi
	Compression: 35,000 psi,
	Shear: 35,000 psi
Yield point tension psi	@ 35,000 psi.
Modulus of elasticity, tension or compression, psi	10,000,000
Modulus of elasticity, shear, and psi	@ 3,750,000
Weight per cu in., lb:	0.10

talking about—and to understand its significance. Remember Voltaire: "If you wish to converse with me, define your terms."

Let's take a look at a few other engineering terms and their definitions, so that we will be able to converse. Many of the following engineering terms are from Heisler's *The Wiley Engineer's Desk Reference: A Concise Guide for the Professional Engineer* (1984) and Giachino and Weeks' *Welding Skills* (1985)—which should be standard reference texts for any safety engineer.

Stress—the internal resistance a material offers to being deformed. Measured in terms of the applied load over the area (see Figure 6.8).

Strain—the deformation that results from a stress. Expressed in terms of the amount of deformation per inch.

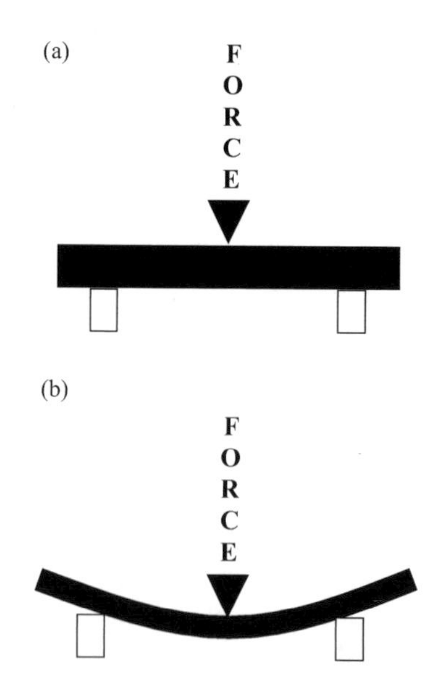

FIGURE 6.8 (a) Stress—measured in terms of the applied load over the area; (b) Strain—expressed in terms of amount per square inch.

Intensity of stress—the stress per unit area, usually expressed in pounds per square inch. Due to a force of **P** pounds producing tension, compression, or shear on an area of **A** square inches, over which it is uniformly distributed. The simple term, stress, is normally used to indicate intensity of stress.

Ultimate stress—the greatest stress that can be produced in a body before rupture occurs.

Allowable stress or working stress—the intensity of stress that the material of a structure or a machine is designed to resist.

Elastic limit—the maximum intensity of stress to which a material may be subjected and return to its original shape upon the removal of stress (see Figure 6.9).

Yield point—the intensity of stress beyond which the change in length increases rapidly with little (if any) increase in stress.

Modulus of elasticity—the ratio of stress to strain, for stresses below the elastic limit. By checking the modulus of elasticity, the comparative stiffness of different materials can readily be ascertained. Rigidity and stiffness is very important for many machine and structural applications.

Poisson's ratio—the ratio of the relative change of diameter of a bar to its unit change of length under an axial load that does not stress it beyond the elastic limit.

Intensity of stress—the stress per unit area, usually expressed in pounds per square inch. Due to a force of **P** pounds producing tension, compression, or shear on an area of **A** square inches, over which it is uniformly distributed. The simple term, stress, is normally used to indicate intensity of stress.

Tensile strength—the property that resists forces acting to pull the metal apart—a very important factor in the evaluation of a metal (see Figure 6.10).

Compressive strength—the ability of a material to resist being crushed (see Figure 6.11).

Bending strength—that quality that resists forces from causing a member to bend or deflect in the direction in which the load is applied—actually a combination of tensile and compressive stresses (see Figure 6.12).

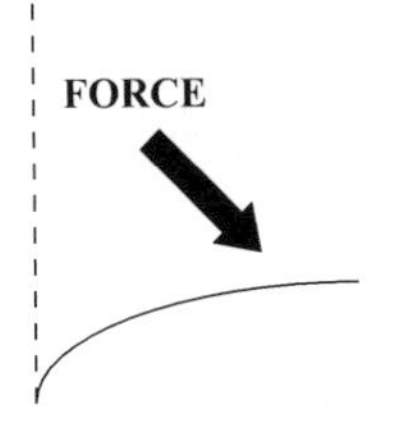

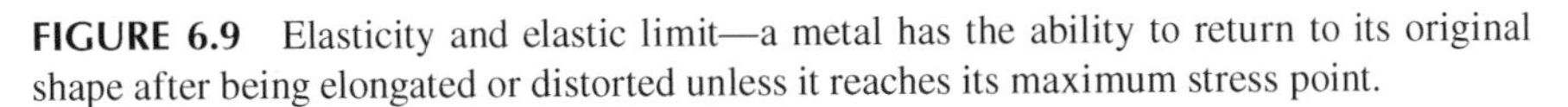

FIGURE 6.9 Elasticity and elastic limit—a metal has the ability to return to its original shape after being elongated or distorted unless it reaches its maximum stress point.

FIGURE 6.10 A metal with tensile strength resists pulling forces.

FIGURE 6.11 Compressive strength—a metal's ability to resist crushing forces.

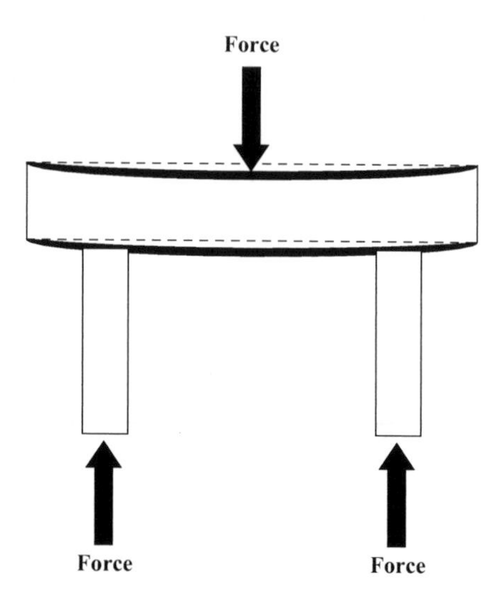

FIGURE 6.12 Bending strength (stress)—a combination of tensile strength and compression stresses.

FIGURE 6.13 Torsional strength—a metal's ability to withstand twisting forces.

Torsional strength—the ability of a metal to withstand forces that cause a member to twist (see Figure 6.13).

Shear strength—how well a member can withstand two equal forces acting in opposite directions (see Figure 6.14).

Fatigue strength—the property of a material to resist various kinds of rapidly alternating stresses.

Impact strength—the ability of a metal to resist loads that are applied suddenly and often at high velocity.

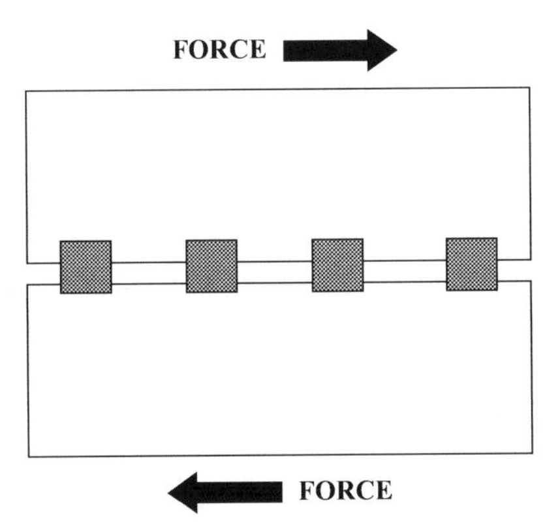

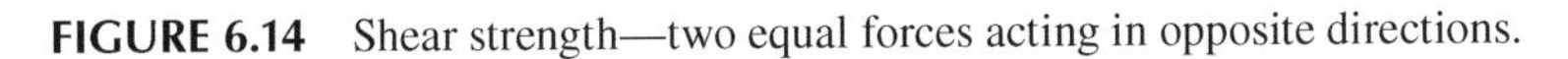

FIGURE 6.14 Shear strength—two equal forces acting in opposite directions.

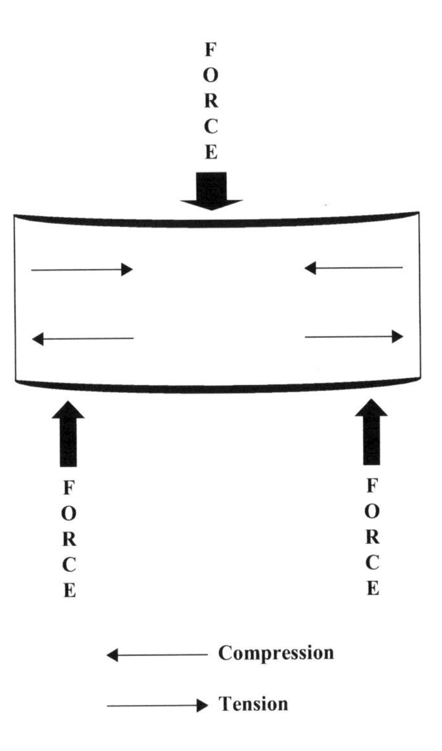

FIGURE 6.15 Distribution of stress in a beam cross section during bending.

Ductility—the ability of a metal to stretch, bend, or twist without breaking or cracking (see Figure 6.15).

Hardness—the property in steel that resists indentation or penetration.

Brittleness—a condition whereby a metal will easily fracture under low stress.

Toughness—may be considered as strength, together with ductility. A tough material can absorb large amounts of energy without breaking.

Malleability—the ability of a metal to be deformed by compression forces without developing defects, such as encountered in rolling, pressing, or forging.

Friction

Earlier, in discussing the principle of the inclined plane, we ignored the effect of friction. In actual use, friction cannot be ignored and you must have some understanding of its characteristics and applications. Friction deals with one body in contact with another that is on the verge of sliding or is sliding. Friction allows us to walk, ski, drive vehicles, and power machines (among other things). Whenever one body slides over another, frictional forces opposing the motion are developed between them. Friction force is the force tangent to the contact surface that resists motion. If motion occurs, the resistance is due to kinetic friction, which is normally lower than the value for static friction. Contrary to common perception, the degree of smoothness of a surface area is not responsible for these frictional forces; instead, the molecular structure of the materials is responsible. The coefficient of friction M (which differs among different materials) is the ratio of the frictional force F to the normal force N between two bodies.

$$M = \frac{F}{N} \tag{6.4}$$

For dry surfaces, the coefficient of friction remains constant, even if the weight of an object (i.e., the force N) is changed. The force of friction (F) required to move the block changes proportionally. Note that the coefficient of friction is independent of the area of contact, which means that pushing a brick across the floor requires the same amount of force, whether it is on end, on edge, or flat. The coefficient of friction is useful in determining the force necessary to do a certain amount of work. Temperature changes only slightly affect friction. Friction causes wear. To overcome this wear problem (to reduce friction), lubricants are used.

Pressure

Pressure, in mechanics, is defined as the force per unit of area or

$$\text{Pressure} = \frac{\text{Total force}}{\text{area}} \tag{6.5}$$

Pressure is usually expressed in terms of force per unit of area, as in pounds per square inch when dealing with gases, or in pounds per square foot when dealing with weight on a given floor area. The pressure exerted on a surface is the perpendicular force per unit area that acts upon it. Gauge pressure is the difference between total pressure and atmospheric pressure.

Specific Gravity

Specific gravity is the ratio of the weight of a substance to the weight of an equal volume of water, a number that can be determined by dividing the weight of a body by the weight of an equal volume of water. Since the weight of any body per unit of volume is called density, then:

$$\text{Specific gravity} = \frac{\text{density of body}}{\text{density of water}} \tag{6.6}$$

Example: The density of a particular material is 0.24 pounds per cubic inch, and the density of water per cubic inch is 0.0361 pounds per cubic inch.

$$\text{Then : Specific gravity of the material} = \frac{0.24}{0.0361} = 6.6$$

The material is 6.6 times as heavy as water. This ratio does not change, regardless of the units that may be used, which is an advantage for two reasons: (1) it will always be the same for the same material, and (2) it is less confusing than the term density, which changes as the units change.

Force, Mass, and Acceleration

According to Newton's second law of motion:

The acceleration produced by an unbalanced force acting on a mass is directly proportional to the unbalanced force, in the direction of the unbalanced force, and inversely proportional to the total mass being accelerated by the unbalanced force.

If we express Newton's second law mathematically, it is greatly simplified and becomes

$$F = ma \tag{6.7}$$

This equation is extremely important in physics and engineering. It simply relates acceleration to force and mass. Acceleration is defined as the change in velocity divided by the time taken. This definition tells us how to measure acceleration. $F = ma$ tells us what causes the acceleration—an unbalanced force. Mass may be defined as the quotient obtained by dividing the weight of a body by the acceleration caused by gravity. Since gravity is always present, we can, for practical purposes, think of mass in terms of weight, making the necessary allowance for gravitational acceleration.

Centrifugal and Centripetal Forces

Two terms the engineer must be familiar with are centrifugal and centripetal forces. Centrifugal force is a concept based on an apparent (but not real) force. It may be regarded as a force that acts radially outward from a spinning or orbiting object (a ball tied to a string whirling about), thus balancing a real force, the centripetal force (the force that acts radially inward).

This concept is important in vehicle-on-wheels engineering because many of the machines encountered on the job, during assembly or while working on a vehicle with wheels, may involve rapidly revolving wheels or flywheels. If the wheel is revolving fast enough, and if the molecular structure of the wheel is not strong enough to overcome the centrifugal force, it may fracture. Pieces (shrapnel) of the wheel would fly off tangent to the arc described by the wheel. The safety implications are obvious. Any worker using such a device, or near it may be severely injured when the rotating member ruptures. This is what happens when a grinding wheel on a pedestal grinder "bursts." Rim speed determines the centrifugal force, and rim speed involves both the speed (rpm) of the wheel and the diameter of the wheel.

Stress and Strain

In materials, stress is a measure of the deforming force applied to a body. Strain (which is often erroneously used as a synonym for stress) is really the resulting change in its shape (deformation). For perfectly elastic material, stress is proportional to stain. This relationship is explained by Hooke's Law, which states that the deformation of a body is proportional to the magnitude of the deforming force, provided that the body's elastic limit is not exceeded. If the elastic limit is not reached, the body will return to its original size once the force is removed. For example, if a spring is stretched by 2 cm by a weight of 1 N, it will be stretched by 4 cm by a weight of 2 N, and so on; however, once the load exceeds the elastic limit for the spring, Hooke's Law will no longer be obeyed, and each successive increase in weight will result in a greater extension until the spring finally breaks. Stress forces are categorized in three ways:

1. Tension (or tensile stress), in which equal and opposite forces that act away from each other are applied to a body; tends to elongate a body.
2. Compression stress, in which equal and opposite forces that act toward each other are applied to a body; tends to shorten it.
3. Shear stress, in which equal and opposite forces that do not act along the same line of action or plane are applied to a body; tends to change its shape without changing its volume.

PRINCIPLES OF MECHANICS

In this section, we discuss mechanical principles: statics, dynamics, beams, floors, columns, electric circuits, and machines. The vehicle-on-wheels engineer should have at least some familiarity with all of these. **Note**: safety engineers whose functions are to verify design specifications (with safety in mind) should have more than just a familiarity with these topics.

Statics

Statics is the branch of mechanics concerned with the behavior of bodies at rest and forces in equilibrium and distinguished from dynamics (concerned with the behavior of bodies in motion). Forces acting on statics do not create motion. Static applications are bolts, welds, rivets, load-carrying components (ropes and chains), and other structural elements. A common example of a static situation is in a bolt and plate assembly. The bolt is loaded in tension and holds two elements together.

Welds

Welding is a method of joining metals to achieve a more efficient use of the materials and faster fabrication and erection. Welding also permits the designer to develop and use new and aesthetically appealing designs and saves weight because connecting plates are not needed and allowances need not be made for reduced load-carrying ability due to holes for rivets, bolts, and so on (Heisler, 1984). Simply put, the welding process joins two pieces of metal together by establishing a metallurgical bond between them. Most processes use a fusion technique; the two most widely used are arc welding and gas welding.

In the welding process, where two pieces of metal are joined together, the mechanical properties of metals are important, of course. The mechanical properties of metals primarily measure how materials behave under applied loads—in other words, how strong a metal is when it comes in contact with one or more forces. The important point is that if you know and use the strength properties of a metal, you can build a structure that is both safe and sound.

In welding, the welder must know the strength of his weld as compared with the base metal to produce a weldment that is strong enough to do the job. Thus, the welder is just as concerned with the mechanical properties of metals as is the engineer.

Dynamics

Dynamics (kinetics in mechanics) is the mathematical and physical study of the behavior of bodies under the action of forces that produce changes of motion in them. In dynamics, certain properties are important: center of gravity, displacement, velocity, acceleration, momentum, kinetic energy, potential energy, work, and power. Vehicle-with-wheels engineers work with these properties to determine, for example, if rotating equipment will fly apart and cause injury to workers, or to determine the distance needed to stop a vehicle in motion.

Hydraulics and Pneumatics—Fluid Mechanics

Hydraulics (liquids only) and pneumatics (gases only) make up the study of fluid mechanics, which in turn is the study of forces acting on fluids (liquids and gases are considered fluids). Engineers encounter many fluid mechanics problems and applications of fluid mechanics. In particular, engineers working in chemical industries, or in or around processes using or producing chemicals need an understanding of flowing liquids or gases to be able to predict and control their behavior.

REFERENCES

ASSE (1988). The American Society of Safety Engineers MSDS hyperglossar. Accessed 11/30/21 @ www.ilpi.com.msds/ref/asse.

Heisler, S. I., *The Wiley Engineer's Desk Reference*. New York: John Wiley and Sons, 1984.

7 Dynamics of Vehicle Motion

In the engineering and design of vehicles-on-wheels vehicle dynamics is the study of the vehicle in motion and how it behaves in motion. It is important for design and engineering professionals to fully understand what a vehicle does and basically what does it add up to. Essentially, EVs work in generally the same way as ones powered by gas, diesel, biodiesel, and hydrogen. There is a fuel source, a drive unit, and a gearbox to provide forward and backward motion. In short, a vehicle-on-wheels includes subsystems/modules as follows.

- **EV power module**—includes EV Traction, Electronic Control Unit (ECU), Transmission Control Unit (TCU), motor, gear box—single-speed transmission, drive axles.
- **EV chassis module**—includes suspension, steering, braking and parking, tires and wheels.
- **EV body module**—includes bonnet, doors, roof, trims, and so forth.

Let's get back to vehicle dynamics (aka vehicle mechanics—vehicle dynamics is a subset of engineering dealing with and based on established mechanics). Simply, vehicle dynamics for vehicles-on-wheels is the study of vehicle motion. More specifically, vehicle dynamics is the study of how a vehicle's forward movement changes in response to driver inputs, propulsion system outputs, ambient conditions, air/surface/water conditions, and so forth. Simply, and to the point vehicle dynamics is all about the study of how much energy is required to propel vehicles-on-wheels and the limiting factors involved in determining the amount of energy required to move the vehicle.

Again, it is all about motion. Okay, that is logical but what are the factors affecting vehicle dynamics?

The factors affecting vehicle dynamics are many and varied, including drivetrain and braking, suspension and steering, distribution of mass, aerodynamics, and tires. In the vehicles-on-wheels application the fundamentals of vehicle design are paramount and embedded in basic physics (mechanics of). With regard to pure science impact vehicle dynamics is all about Newton's second law of motion that states *the acceleration of an object is proportional to the net force exerted on it*. Okay, in this presentation the term "net force" refers to the amount of force acting on the vehicle-on-wheels.

FACTORS AFFECTING VEHICLE DYNAMICS

As mentioned, the factors affecting vehicle dynamics are many and varied, including drivetrain and braking, suspension and steering, distribution of mass, aerodynamics, and tires. The dynamics that govern vehicle motion based on the forces acting on a

DOI: 10.1201/9781003332992-7

rolling vehicle include aerodynamic drag, rolling resistance, hill climbing, linear and angular acceleration, and tractive force. Let's look at each of the factors affecting vehicle dynamics.

Vehicle-on-Wheels Layout

In the drivetrain and braking factor category of vehicle dynamics vehicle-on-wheels layout is derived from the location of the engine and the drive wheels. The layouts can be divided into the three categories of front-wheel drive (FWD), rear-wheel drive (RWD), and four-wheel drive (4WD). In practice the many different combinations of engine location and driven wheels actually employed are dependent on the application for which the vehicle-on-wheels will be used.

Powertrain

Simply, the powertrain consists of the units that provide power to the wheels of the vehicle.

Braking System

The braking system is composed of mechanical devices that restrain motion by absorbing energy from the moving system.

Geometry of Suspension, Steering, and Tires

- One of the considerations in suspension and steering systems in vehicles-on-wheels is steering geometry, namely the *Ackermann steering geometry*; it is a consideration that is focused on the geometric arrangement of linkages used to steer the vehicle-on-wheels and the intention is to avoid the need for tires to slip sideways when following whatever the path around a curve. The geometric solution is for all wheels to have their axles arranged as radii of circle with a common center point (Norris, 1906). Okay, what does all this mean? It means that in "turntable steering" as the rear wheels are fixed, the center point must be on line extended from the rear axle. Note that intersecting the axes of the front wheels on this line as well requires that the inside front wheel be turned, when steering, through a greater angle than the outside wheel (Norris, 1906).

  Instead of the preceding turntable steering, where both front wheels turned around a common pivot, each wheel gained its own pivot, close to its own hub. While more complex, this arrangement enhances controllability by avoiding large inputs from road surface variations being applied to the end of a long lever arm, as well as greatly reducing the fore-and-aft travel of the steered wheels (Norris, 1906).

  Note that modern vehicles-on-wheels do not use Ackerman steering, even though the principle is good for slow-speed maneuvers but it ignores important dynamic and complaint effects.

- *Axle track* in vehicles-on-wheels refers to those vehicles having two wheels on an axle; it is the distance between the hub flanges on an axle (Car Handling Basics, 2022). Track refers to the distance between the centerline of two wheels on the same axle. Axle and track are commonly measured in millimeters or inches (BMW M3 E46, 2022).
- *Camber angle* is one of the angles made by the wheels of a vehicle; stated differently, it is the angle between the vertical axis of a wheel and the vertical axis of the vehicle when viewed from the front or rear.
- *Caster angle* causes a wheel to align with the direction of travel. Caster displacement moves the steering axis ahead of the axis of rotation.
- *Ride height* or *ground clearance* is the distance or space between the base of the vehicle tire and the lowest point of the vehicle (usually the axle).
- *Roll center* of a vehicle is the notional point at which the cornering forces in the suspension are reacted to the vehicle body.
- *Scrub radius* is the distance at the road surface between the tire center line and the steering axis inclination.
- *Steering ratio* is the ratio between the turn of the steering wheel (in degrees) or handlebars and the turn of the wheels (in degrees). For electric motorcycles and bicycles the steering ratio is 1:1, because the steering wheel is attached to the front wheel. In most electric passenger cars the ratio is 12:1 and 20:1 (ratios 13–14 are considered fast and ratios above 18 are considered slow).
- *Toe*, in vehicles-on-wheels toe (aka *tracking*), as a function of static geometry, and kinematic and compliant effects is the symmetric angle that each wheel makes with the longitudinal axis of the vehicle.
- *Wheel alignment* sometimes referred to as breaking or tracking consists of adjusting angles of wheels to manufacturer specifications.
- *Wheelbase* is the distance between the front and rear axles of a vehicle-on-wheels.

Some aspects of vehicle dynamics are due to mass and its distribution. Mass distribution is the spatial distribution of mass within a solid body. These include:

- *Center of mass* is the unique point where the weight relative portion of the distribution sums to zero.
- *Moment of inertia* depends on the moment of the very different moment of inertia depending on the location and orientation of the axis or rotation.
- *Roll moment* is a product of force and distance and causes a vehicle to roll, rotating about its longitudinal axis.
- *Sprung mass* in a vehicle-on-wheels with a suspension is the portion of the vehicle's total mass that is supported by the suspension, including in most applications approximately half of the suspension itself.
- *Unsprung mass* sometimes called unsprung weight of a vehicle is the mass of the suspension directly connected.
- *Weight distribution* is the apportioning of weight with a vehicle-on-wheels.

Some aspects of vehicle dynamics are due to aerodynamics aspects. These include:

- *Automobile drag coefficient* is a common measure in automotive design. Drag is the force that acts parallel to and in the same direction as the airflow.
- *Automotive aerodynamics* is the study involved with reducing drag, wind noise, and preventing undesirable lift in vehicles-on-wheels.
- *Center of pressure* is the point where the total sum of a pressure field acts on a body causing a force to act through that point.
- *Downforce* is a downward lift force created by the aerodynamic features of a vehicle-on-wheels.
- *Ground effect (automobiles)* is a series of effects that have been exploited in automotive aerodynamics to create downforce.

Vehicle dynamics is directly affected by the tires of a vehicle-on-wheels. For instance, one of the interesting factors affecting vehicle dynamics is known as the *Magic Formula* tire models. Developed by Hans Pacejka the Magic Formula actually consists of a series of formulae that Pacejka developed over the last 20 years.

The significance of the Magic Formula?

Well, the truth be told, the Magic Formula is widely used in professional vehicle dynamics simulations because they are reasonably accurate, very easy to program, and more importantly they solve quickly.

Used for what?

Good question. The Magic Formula is, as previously stated, a series of tire design models that Pacejka developed over time.

So, what is so magical about the Magic Formula?

First off, there is no particular physical basis for the structure of the equations chosen, but they fit a wide variety of tire constructions and operating conditions—in short, the Magic Formula is not only easy to use but is adaptable to several different applications or requirements and is widely used in professional vehicle dynamics simulations and they are reasonably accurate, easy to program, and solve quickly (Plasterk, 1989).

Along with the Magic Formula there are other aspects of vehicle dynamics related to tires include camber thrust, circle of forces (i.e., a useful way of thinking about the dynamic interactions between the vehicle's tire and road), contact patch (i.e., the pneumatic touch of the tire to road surface), cornering force, ground pressure, pneumatic trail (i.e., the trail of the tire), and radial force variation (i.e., road force variation is the property of a tire that affects steering, traction, braking, and load support) (Cortez, 2014). A property of pneumatic tires that describes the delay between when a slip angle is introduced and when cornering force reaches its steady-state value is known as *relaxation length* (Pacejka, 2005). *Rolling resistance* is the force resisting the motion when a body (tire) rolls on a surface. The torque a tire develops as it rolls along is known as *self-aligning torque (aka aligning torque, aligning moment, SAT, or MS)*; it tends to steer it, that is, rotate it around its vertical axis. *Skid* occurs when one or two tires slip relative to the road. *Slip angle or sideslip* is the angle between the direction in which a wheel is pointing and the direction in which it is actually traveling (Pacejka, 2005). Another characteristic in vehicle dynamics related to tires is *slip*, which is the relative motion between the tire and the road surface it is moving on. A subset of slip is *spinout* that occurs when a vehicle rotates in one direction during a skid. *Steering ratio* refers to the ratio between the turn of the steering wheel (in degrees) and the turn of the wheel (in degrees) (Pacejka, 2005). The behavior of tires under load is called *tire load sensitivity.*

Pure Dynamics

Beyond distribution of mass, aerodynamics, and tires some attributes and aspects of vehicle dynamics are purely dynamic. For example, *body flex* is a lack of rigidity in a vehicle-on-wheels' chassis. The axial rotation of a vehicle-on-wheels toward the outside of a turn called *body roll* is another example of a purely dynamic aspect of vehicle dynamics. Another term used in vehicle dynamics for a pure dynamic attribute or aspect is *bump steer.* Bump steer is caused when one wheel falls down into a rut or hole or hits a bump causing the vehicle-on-wheels to turn itself.

Note that there are several other factors, attributes, and aspects of pure dynamics that impact vehicle dynamics and more importantly vehicle operation. A few of these other factors, attributes, and aspects of aerodynamics affecting operation of vehicles-on-wheels include directional stability, critical speed, pitch, yaw, roll, speed wobble, understeer, oversteer, weight transfer, and yaw—it is these factors that affect vehicle operation.

However, if we have to sum up the factors, attributes, and aspects of vehicle dynamics related to environmental impact on vehicles-on-wheels it comes down to the forces acting on a rolling vehicle-on-wheels. And these forces are all about aerodynamics; at least, aerodynamics is where we begin—and we do this by describing aerodynamic drag, rolling resistance, linear acceleration, hill climbing, angular acceleration, and the total of them all, and this equates to the total tractive force required to propel the vehicle.

AERODYNAMICS[1]

It is interesting that whenever the term "aerodynamics" is mentioned anywhere at any time it is likely that those exposed to the term are apt to equate the term with aircraft. And this makes sense when aerodynamics is thought of and defined in its simplest terms; that is, it is the way air moves around things and aircraft certainly depends on the rules of aerodynamics to enable then to fly. However, it is important to point out that anything that moves through air reacts to aerodynamics. A kite in the sky reacts to aerodynamics and aerodynamics even acts on vehicles-on-wheels because air flows around cars and trucks and buses and other moving objects.

To illustrate the importance of aerodynamics and to provide foundational information for the impact on electric vehicles-on-wheels we begin by addressing the four forces of flight—lift, weight, thrust, and drag. These forces make an object move up and down, and faster or slower. How much of each force is present changes how the

[1] Much of the information in this section is adapted from NASA (2017) Aerodynamics. Accessed 6/1/22 @ https://www.nasa.gov/offices.

object moves through the air. Let's take a closer look at each of these four forces: weight, lift, thrust, and drag.

Weight

Everything around us on earth has *weight*. This force is the result of gravity pulling down on objects. To fly, an aircraft needs something to push it in the opposite direction from gravity. The weight of an object controls how strong the push has to be. Flying a kite is a lot easier than pushing upward on a jumbo aircraft.

Lift

With regard to the push needed to move aircraft and kites upward the push needed is called *lift*. Simply, lift is the force that is the opposite of weight. Again, in regard to aircraft everything that flies must have lift. Again, for an aircraft to move upward, it must have more lift than weight. It is all about light air, well, with hot air balloons they have lift because the hot air inside is lighter than the air around it. The hot air rises and takes balloons with it. Helicopter's lift comes from the rotor blades at the top of the helicopter. The helicopter is raised in the air via its motion in the air. However, lift for an aircraft is all about its wings.

So, the question is: how does an aircraft's wings provide lift?

The answer is: It's all about the shape of the aircraft's wings that provides the lift. Airplanes' wings are curved on top and flatter on the bottom. The shape makes the air travel over the top faster than under the bottom. Consequently, less air pressure is on top of the wing. This condition makes the wing, and the airplane it's attached to, move up. Using curves to change air pressure is a trick used on many aircrafts. Helicopter blades, kites, and sailboats use this trick.

Drag

Before we get into drag and its effect on electric vehicles-on-wheels we need to get down a few definitions pertinent to the drag effect. First we need to define drag coefficient (Cd) that is a dimensionless quantity that is used to quantify the drag or resistance of an object in a fluid environment, such as air—air is indeed classified as a fluid. Second, when addressed as vehicles-on-wheels, even though what is discussed and detailed could be applied to trains on rails, it is the vehicles that use the highways, streets, race tracks, or farm fields that is what really is being addressed.

Okay, now for drag force, this is best addressed and explained via Equation (7.1).

$$\text{Fd} = 1/2\text{pu}2\text{ACd} \tag{7.1}$$

where

Fd = drag force, which is a definition in the direction of flow

p = mass density of the fluid

u = flow velocity relative to the object
A = the reference area
Cd = drag coefficient (dimensionless)

Note: If the fluid is a gas (air) Cd depends on both the Reynolds Number (Re) and the Mach Number. Basically, the *Reynolds Number (Re)* helps predict flow pattern in different fluid flow (air flow) situations. Low Re equals laminar flow (sheet-like flow), while at high Re flows tend to be turbulent. Mach Number (M or Ma) is a dimensional quantity in fluid dynamics—the ratio of flow velocity past a boundary to the local speed of sound.

Note that the shape of the object has effects on the amount of drag.

In simple terms drag can be defined as a force that tries to slow somethings down. Basically, it makes it hard for an object to move. Keep in mind that it is harder to run through water than through air. That is because water causes more drag than air. The shape of an object also changes the amount of drag as shown in Figure 7.1. Most round surfaces have less drag than flat ones. Narrow surfaces usually have less drag than wide ones. Simply, the more air that strikes a surface, the more drag it makes.

Thrust

The force that is the opposite of drag is *thrust*. Thrust is the push that moves electric vehicles-on-wheels forward.

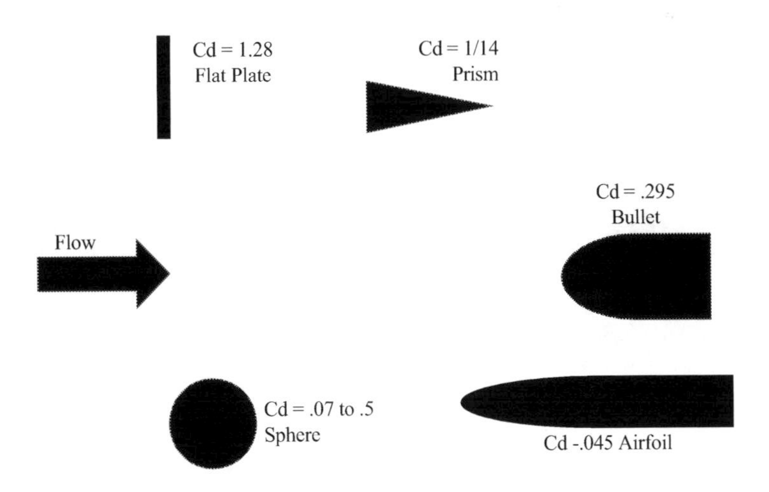

FIGURE 7.1 Shape effects on drag—all objects have the same frontal area. (NASA accessed 05/31/22@ www.grc.nasa.gov/www/Ke12/airplane/shapter.html.)

Rolling Resistance

Rolling resistance (aka rolling friction or rolling drag) is the force that resists the motion of a body rolling on a surface (see Figure 7.2). The rolling resistance can be expressed in various ways but for our purposes the generic equation is shown here.

$$F_r = c \cdot W \tag{7.2}$$

where

F_r = rolling resistance or rolling fiction or rolling drag (N, lb_f)
c = rolling resistance coefficient—dimensionless—rolling resistance coefficient, c, is influenced by different variables like wheel design, rolling surface, wheel dimensions, and more (see Table 7.1)
$W = m\, a_g$
m = mass of the body (kg, lb)
a_g = acceleration due to gravity (9.81 m/s², 32.174 ft/s²)

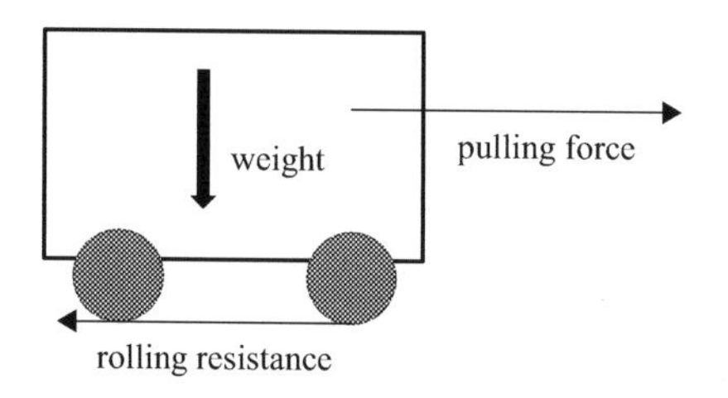

FIGURE 7.2 Rolling resistance.

TABLE 7.1
Some Typical Rolling Friction Coefficients

Rolling Resistance Coefficient, c	Typical Vehicle Tire Surface Conditions
0.006–0.01	Truck tire on asphalt
0.01–0.015	Ordinary car tires on concrete, asphalt, small cobbles
0.02	Car tires on tar or asphalt
0.02	Car tires on rolled gravel
0.03	Car tires on large, worn cobble
0.04–0.08	Car tire on solid sand, loose gravel, hard sand
0.2–0.4	Car tire on loose sand

DID YOU KNOW?

Note that it takes most of the usable energy to overcome drag and rolling resistance: aerodynamics uses up to 53% of the energy, while drive train uses about 6% and auxiliary equipment uses roughly 9% and rolling resistance uses 30+% of usable energy—the good news is these losses can be reduced using presently available technology.

Hill Climbing

A vehicle-on-wheels traveling on a flat service at constant velocity experiences two major forces opposing the vehicle, aerodynamic drag and rolling resistance. However, if that same vehicle-on-wheels is traveling up a hill, we also need to account for gravity. One of the commonly used equations to calculate the force needed to travel up a hill is Equation (7.3).

$$F_{uh} = mg\sin\psi \tag{7.3}$$

where

F_{uh} = force required to travel uphill
mg = m is the mass of the vehicle-on-wells (in kg), g is gravity (9.81 m/s^2)
Ψ = is the vertical angle of the road relative to flat (in radians)

Figure 7.3 illustrates why sine is the appropriate function here.

Linear Acceleration

Linear acceleration is a term related to an object in movement. Acceleration is the measure of how quickly the velocity of any moving object changes. Therefore, the acceleration is the change in the velocity, divided by the time. Acceleration has both magnitude and direction like a vector where both velocity and force are vector quantities.

Okay, acceleration and velocity have been, to a point, explained, now it is time to define linear acceleration which we begin by stating that any object moving in a straight line will be accelerating if its velocity is increasing or decreasing during a

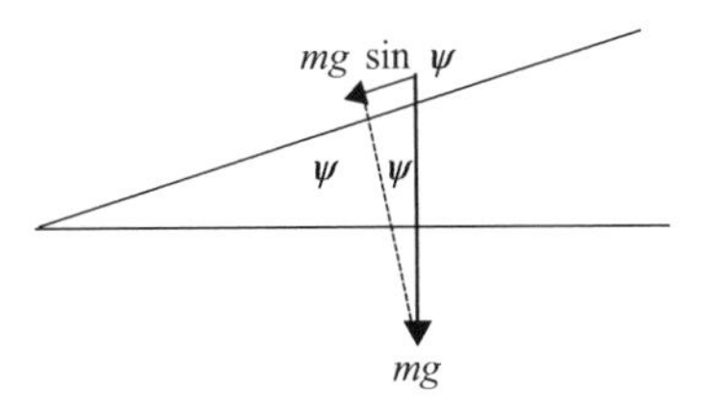

FIGURE 7.3 As shown here hill climbing force is equal to mg sin Ψ.

given period of time. Acceleration can be either positive or negative depending on whether the velocity is increasing or decreasing. Thus, the acceleration is described as the rate of change of velocity of an object. Acceleration is a vector quantity that is described as the frequency at which the object's velocity changes. So, we can say in simple terms that

$$\text{Linear Acceleration} = \frac{\text{Change in Velocity}}{\text{Time Taken}} \tag{7.4}$$

and its unit is meters per second squared or m s^{-2}.

Angular Acceleration

Angular acceleration refers to the time rate of change of angular velocity and is measured in units of angle per unit time square (which in SI units is radians per second squared) and is usually represented by the symbol alpha ($\boldsymbol{\alpha}$).

Tractive Force

Tractive force is simply the sum of all the forces we have discussed—it is the propulsion unit from an electric vehicle-on-wheels. As used in this text the term tractive force refers to total traction a vehicle exerts on a surface.

REFERENCES

BMW M3 E46. Car.info. Accessed 5/24/22 @ https://www.bmusa.com.

Car Handling Basics (2022). How to and Design Tips. Accessed 5/24/22 @ https://www.build-yourown race car.com/racecar-handling-basics-and-design/.

Cortez, P. (2014). Repairing persistent pulls, drifts, shimmies & vibration. Accessed 5/28/22 @ https://classic.artsautomotive.xom/GSP9700.htm.

Norris, W. (1906), Steering. Accessed 5/24/22 @ https://archive.org/details/modern steam roadw00norrich.

Pacejka, H. B. (2005). *Tyre and Vehicle Dynamics*, 2nd ed. Accessed 5/28/22 @ https://www.SAE.org, SAE International, p. 22.

Plasterk, K. J. (1989). The End of the First Era.: A Farewell to Hans Pacejka. Accessed 27 May, 2022 @ https://www.woldcat.org/issn/oo42-3114.

8 Electric Motors

MOTORS

A variety of different types of electric motors and other important components make up the power train that provides battery-stored electrical energy that is converted to rotational power that is required to move an electric vehicle-on-wheels (see Figure 8.1). Of course, electric motors are used for other purposes than powering EVs. For instance, at least 60% of the electrical power fed to a typical waterworks and/or wastewater treatment plant is consumed by electric motors. One thing is certain: there is an almost endless variety of tasks that electric motors perform in industry and for personal use. An *electric motor* is a machine used to change electrical energy to mechanical energy to do the work. (Note that a generator does just the opposite; that is, a generator changes mechanical energy to electrical energy.)

Previously, we pointed out that when a current passes through a wire, a magnetic field is produced around the wire. If this magnetic field passes through a stationary magnetic field, the fields either repel or attract, depending on their relative polarity. If both are positive or negative, they repel. If they are of opposite polarity, they attract. Applying this basic information to motor design, an electromagnetic coil, the armature, rotates on a shaft. The armature and shaft assembly are called the rotor. The rotor is assembled between the poles of a permanent magnet and each end of the rotor coil (armature) is connected to a commutator also mounted on the shaft. A commutator is composed of copper segments insulated from the shaft and from each other by an insulting material. As like poles of the electromagnet in the rotating armature pass the stationary permanent magnet poles, they are repelled, continuing the motion. As the opposite poles near each other, they attract, continuing the motion.

As a quick review of importance of the electric motor in an EV, let's again point out that in an electric drive system, an electric motor converts the stored electrical energy in a battery to mechanical energy. Again, an electric motor consists of a rotor (the moving part of the motor) and a stator (the stationary part of the motor). A permanent magnet motor includes a rotor containing a series of magnets and a current-carrying stator (typically taking the form of an iron ring), separated by an air gap.

The technology involved with EVs is dynamic, constantly changing, advancing with the goal of EV manufacturers to improve the motors used to move the vehicles-on-wheels. Much of the current research and development (R&D) is to improve motors in hybrid and plug-in electrical vehicles, with particular focus on improving batteries to increase vehicle range and also on reducing the use of rare earth materials currently used for permanent magnet-based motors.

With regard to rare earth elements (REEs), materials, and metals they are critical to our modern way of life, although few people know or understand this. The truth be told this lack of knowledge or understanding of rare earth elements is

DOI: 10.1201/9781003332992-8

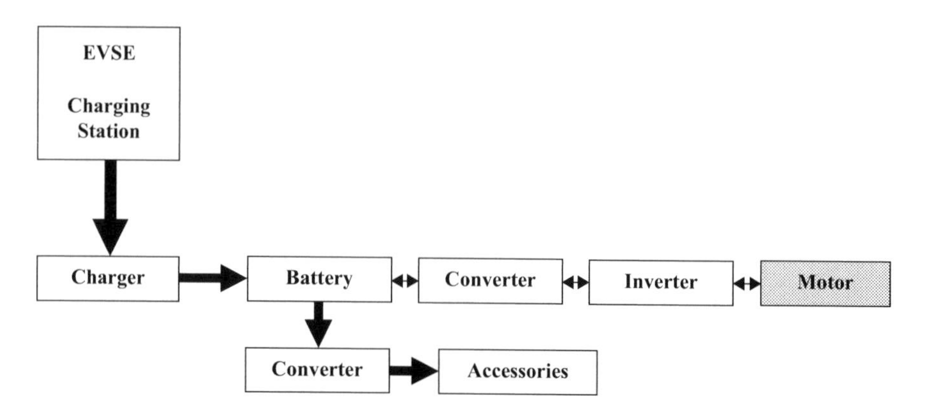

FIGURE 8.1 Electric motor and electric drive system components.

surprising because they are critical ingredients in today's mix of technologies including electronics, electric motors, magnets, batteries, generators, energy storage systems (supercapacitors/pseudocapacitors), emerging applications, and specialty alloys. Rare earth elements are used in various sectors of the US economy including health care, transportation, power generation, petroleum refining, and consumer electronics.

Rare earth materials are extensively used in wind turbine operation to produce electrical power, in some solar applications for electrical power, and in energy storage systems. REEs are also used in electric vehicles, thereby decreasing the need to use fossil fuels for operation.

Concern for the environment and for the impact of environmental pollution has brought about the trend (and the need) to shift from the use and reliance on hydrocarbons to energy power sources that are pollution neutral or near pollution neutral and renewable. We are beginning to realize that we are responsible for much of the environmental degradation of the past and present—all of which is readily apparent today. Moreover, the impact of 200 years of industrialization and surging population growth has far exceeded the future supply of hydrocarbon power sources. So, the implementation of renewable energy sources is surging, and along with it there is a corresponding surge in utilization of rare earth materials for use in energy production.

So, the question is: Why are researchers trying to develop EV motors that do not utilize REEs? It is not because REEs are rare, but because they are not. The problem is source(s); availability of REEs is basically controlled by countries outside the United States. Therefore, not only is the accessibility controlled by entities outside the United States but also the cost involved—economics is always in play in the manufacturing businesses. The United States possesses locations within this country to obtain REEs but with the exception of one location REEs are not being actively mined.

DC MOTORS

The construction of a DC motor is essentially the same as that of a DC generator. However, it is important to remember that the DC generator converts mechanical energy into electrical energy and back into mechanical energy. A DC generator may be made to function as a motor by applying a suitable source of DC voltage across the normal output electrical terminals. There are various types of DC motors, depending on the way the field coils are connected. Each has characteristics that are advantageous under given load conditions.

The field coils of *shunt motors* (see Figure 8.2) are connected in parallel with the armature circuit. This type of motor, with constant potential applied, develops variable torque at an essentially constant speed, even under changing load conditions. Such loads are found in machine-shop equipment such as lathes, shapes, drills, milling machines, and so forth.

The field coils of *series motors* (see Figure 8.3) are connected in series with the armature circuit. This type of motor, with constant potential applied, develops variable torque but its speed varies widely under changing load conditions. That is, the speed is low under heavy loads, but becomes excessively high under light loads. Series motors are commonly used to drive electric hoists, winches, cranes, and certain types of vehicles (e.g., electric trucks). In addition, series motors are used extensively to start internal combustion engines.

Compound motors (see Figure 8.4) have one set of field coils in parallel with the armature circuit, and another set of field coils in series with the armature circuit. This type of motor is a compromise between shunt and series motors. It develops an increased starting torque over that of the shunt motor and has less variation in speed than the series motor.

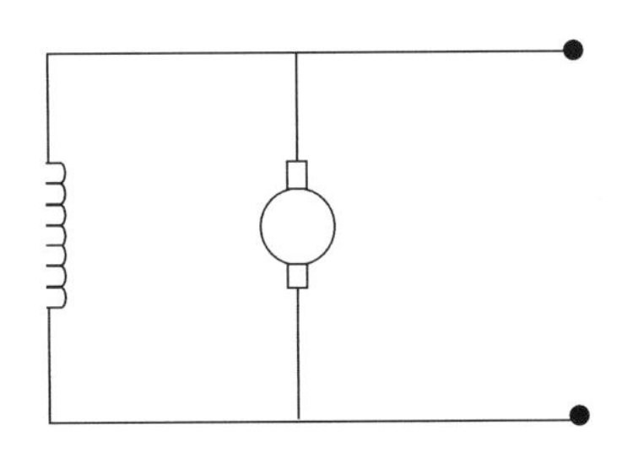

FIGURE 8.2 DC shunt motor.

FIGURE 8.3 DC series motor.

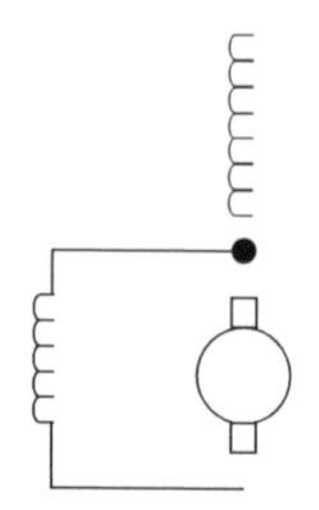

FIGURE 8.4 DC compound motor.

The speed of a DC motor is variable. It is increased or decreased by a rheostat connected in series with the field or in parallel with the rotor. Interchanging either the rotor or field winding connections reverses direction.

AC MOTORS

Alternating current voltage can be easily transformed from low voltages to high voltages or vice versa and can be moved over a much greater distance without too much loss in efficiency. Most of the power generating systems today, therefore, produce alternating current. Thus, it logically follows that a great majority of the electrical motors utilized today are designed to operate on alternating current. However, there are other advantages in the use of AC motors besides the wide availability of AC power. In general, AC motors are less expensive than DC motors. Most types of AC motors do not employ brushes and commutators. This eliminates many problems of maintenance and wear and eliminates dangerous sparking. AC motors are manufactured in many different sizes, shapes, and ratings, for use on an even greater number of jobs. They are designed for use with either polyphase or single-phase power systems. This chapter cannot possibly cover all aspects of the subject of AC motors. Consequently, it will deal mainly with the operating principles of the two most common types—the *induction* and *synchronous motor.*

Induction Motors

The induction motor is the most commonly used type of AC motor because of its simple, rugged construction and good operating characteristics. It consists of two parts: the stator (stationary part) and the rotor (rotating part). The most important type of polyphase induction motor is the *three-phase motor.*

Important Note: A three-phase (3-θ) system is a combination of three single-phase (1-θ) systems. In a 3-θ balanced system, the power comes from an AC generator that produces three separate but equal voltages, each of which is out of phase with the other voltages by 120°. Although 1-θ circuits are widely used in electrical systems, most generation and distribution of AC current is 3-θ.

The driving torque of both DC and AC motors is derived from the reaction of current-carrying conductors in a magnetic field. In the DC motor, the magnetic field is stationary and the armature, with its current-carrying conductors, rotates. The current is supplied to the armature through a commutator and brushes. In *induction motors*, the rotor currents are supplied by electromagnet induction. The stator windings, connected to the AC supply, contain two or more out-of-time-phase currents, which produce corresponding mmfs. These mmfs establish a rotating magnetic field across the air gap. This magnetic field rotates continuously at constant speed regardless of the load on the motor. The stator winding corresponds to the armature winding of a DC motor or to the primary winding of a transformer. The rotor is not connected electrically to the power supply.

The induction motor derives its name from the fact that mutual induction (or transformer action) takes place between the stator and the rotor under operating conditions. The magnetic revolving field produced by the stator cuts across the rotor conductors, inducing a voltage in the conductors. This induced voltage causes rotor current to flow. Hence, motor torque is developed by the interaction of the rotor current and the magnetic revolving field.

With regard to induction motors used in various electric vehicles-on-wheels models the high selling point, significant characteristic is that they have high starting torque and offer high reliability. Note, however, that induction motors have levels of power density and overall efficiency when compared to internal permanent magnetic (IPM) motors (described later). Induction motors are widely available and common in various industries today, including some production vehicles. There is a current problem with induction type motors used in EVs, they are dated—meaning they have been around a long time (aka old hat) and because of their maturity it is unlikely that research could achieve additional (new) improvements in cost, weight, efficiency, and volume for competitive future electric vehicles. Simply, we know induction motors well and that is that.

Synchronous Motors

Okay, let's get back to the basics (that is, to the foundation blocks of basic electric motor principles) and we begin with synchronous motors. Like induction motors, *synchronous motors* have stator windings that produce a rotating magnetic field. However, unlike the induction motor, the synchronous motor requires a separate source of DC from the field. It also requires special starting components. These include a salient-pole field with starting grid winding. The rotor of the conventional type synchronous motor is essentially the same as that of the salient-pole AC generator. The stator windings of induction and synchronous motors are essentially the same.

In operation, the synchronous motor rotor locks into step with the rotating magnetic field and rotates at the same speed. If the rotor is pulled out of step with the rotating stator field, no torque is developed and the motor stops. Since a synchronous motor develops torque only when running at synchronous speed, it is not self-starting and hence needs some device to bring the rotor to synchronous speed. For example, a synchronous motor may be started rotating with a DC motor on a common shaft. After the motor is brought to synchronous speed, AC current is applied to the stator windings. The DC starting motor now acts as a DC generator, which supplies DC field excitation for the rotor. The load then can be coupled to the motor.

Single-Phase Motors

Single-phase (1-θ) motors are so called because their field windings are connected directly to a single-phase source. These motors are used extensively in fractional horsepower sizes in commercial and domestic applications. The advantages of using single-phase motors in small are that they are less expensive to manufacture than other types, and they eliminate the need for 3-phase AC lines. Single-phase motors are used in fans, refrigerators, portable drills, grinders, and so forth.

A single-phase induction motor with only one stator winding and a cage rotor is like a 3-phase induction motor with a cage rotor except that the single-phase motor has no magnetic revolving field at start and hence no starting torque. However, if the rotor is brought up to speed by external means, the induced currents in the rotor will cooperate with the stator currents to produce a revolving field, which causes the rotor to continue to run in the direction, which it was started.

Several methods are used to provide the single-phase induction motor with starting torque. These methods identify the motor as *split-phase*, *capacitor*, *shaded-pole*, and *repulsion-start* induction motor. Another class of single-phase motors is *the AC series* (universal) type. Only the more commonly used types of single-phase motors are described.

Split-Phase Motors

The split-phase motor (see Figure 8.5) has a stator composed of slotted lamination that contains a starting winding and a running winding.

Note: If two stator windings of unequal impedance are spaced 90 electrical degrees apart but connected in parallel to a single-phase source, the field produced will appear to rotate. This is the principle of *phase splitting*.

The starting winding has fewer turns and smaller wire than the running winding, hence has higher resistance and less reactance. The main winding occupies the lower half of the slots and the starting winding occupies the upper half. When the same voltage is applied to both windings, the current in the main winding lags behind the current in the starting winding. The angle θ between the main and starting windings is enough phase difference to provide a weak rotating magnetic field to produce a starting torque. When the motor reaches a predetermined speed, usually 75% of

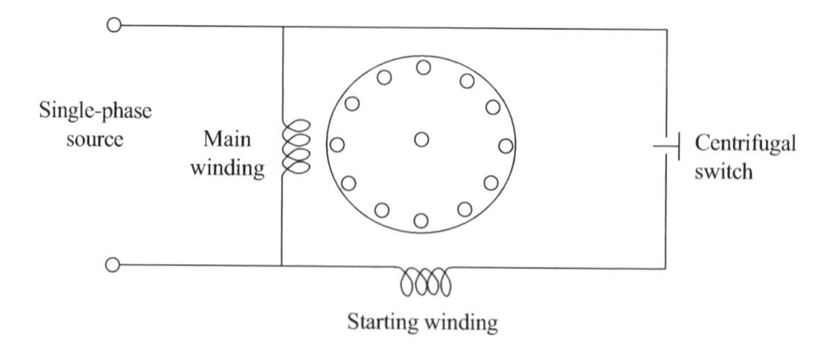

FIGURE 8.5 Split-phase motor.

synchronous speed, a centrifugal switch mounted on the motor shaft opens, thereby disconnecting the starting winding. Because it has a low starting torque, fractional horsepower split-phase motors are used in a variety of equipment such as washers, oil burners, ventilating fans, and woodworking machines. Interchanging the starting winding leads can reverse the direction of rotation of the split-phase motor.

Capacitor Motors

The capacitor motor is a modified form of split-phase motor, having a capacitor in series with the starting winding. The capacitor motor operates with an auxiliary winding and series capacitor permanently connected to the line (see Figure 8.6). The capacitance in series may be of one value for starting and another value for running. As the motor approaches synchronous speed, the centrifugal switch disconnects one section of the capacitor. If the starting winding is cut out after the motor has increased in speed, the motor is called a *capacitor-start motor*. If the starting winding and capacitor are designed to be left in the circuit continuously, the motor is called *capacitor-run motor*. Capacitor motors are used to drive grinders, drill presses, refrigerator compressors, and other loads that require relatively high starting torque. Interchanging the starting winding leads may reverse the direction of rotation of the capacitor motor.

Shaded-Pole Motor

A shaded-pole motor employs a salient-pole stator and a cage rotor. The projecting poles on the stator resemble those of DC machines except that the entire magnetic circuit is laminated and a portion of each pole is split to accommodate a short-circuited coil called a *shading coil* (see Figure 8.7). The coil is usually a single band or strap of copper. The effect of the coil is to produce a small sweeping motion of the field flux from one side of the pole piece to the other as the field pulsates. This slight shift in the magnetic field produces a small starting torque. Thus, shaded-pole motors are self-starting. This motor is generally manufactured in very small sizes, up to 1/20 hp, for driving small fans, small appliances, and clocks.

In operation, during that part of the cycle when the main pole flux is increasing, the shading coil is cut by the flux, and the resulting induced emf and current in the shading coil tend to prevent the flux from rising readily through it. Thus, the greater portion of the flux rises in that portion of the pole that is not near the shading coil.

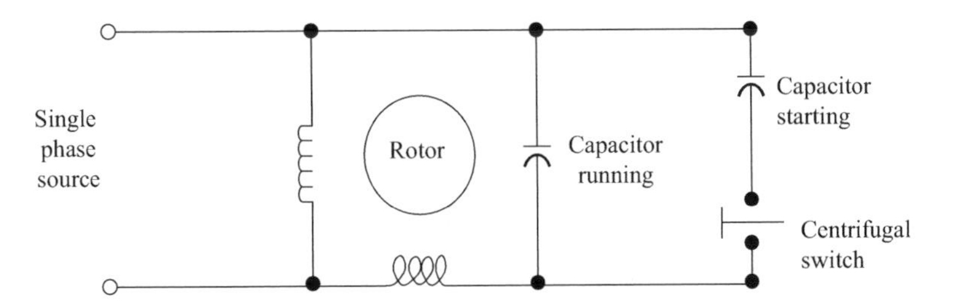

FIGURE 8.6 Capacitor motor.

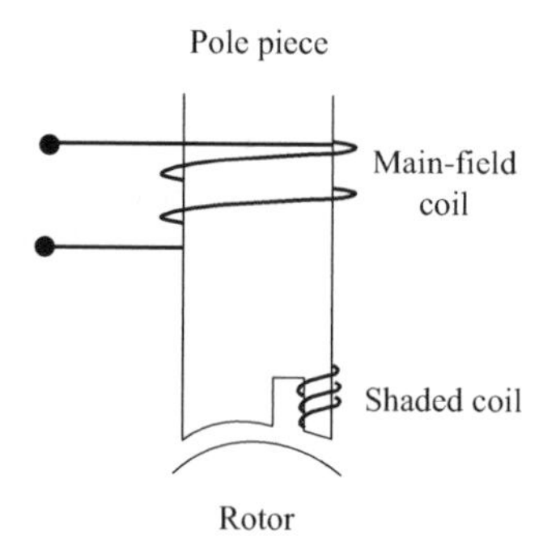

FIGURE 8.7 Shaded pole.

When the flux reaches its maximum value, the rate of change of flux is zero, and the voltage and current in the shading coil are zero. At this time, the flux is distributed more uniformly over the entire pole face. Then as the main flux decreases toward zero, the induced voltage and current in the shading coil reverse their polarity, and the resulting mmf tends to prevent the flux from collapsing through the iron in the region of the shading coil. The result is that the main flux first rises in the unshaded portion of the pole and later in the shaded portion. This action is equivalent to a sweeping movement of the field across the pole face in the direction of the shaded pole. This moving field cuts the rotor conductors and the force exerted on them causes the rotor to turn in the direction of the sweeping field. The shaded-pole method of starting is used in very small motors, up to about 1/25 hp, for driving small fans, small appliances, and clocks.

Repulsion-Start Motor

Like a DC motor, the *repulsion-start motor* has a form-wound rotor with commutator and brushes. The stator is laminated and contains a distributed single-phase winding. In its simplest form, the stator resembles that of the single-phase motor. In addition, the motor has a centrifugal device, which removes the brushes from the commutator and places a short-circuiting ring around the commutator. This action occurs at about 75% of synchronous speed. Thereafter, the motor operates with the characteristics of the single-phase induction motor. This type of motor is made in sizes ranging from 1/2 to 15 hp and is used in applications requiring a high starting torque.

Series Motor

The AC series motor will operate on either AC or DC circuits. When an ordinary DC series motor is connected to an AC supply, the current drawn by the motor is low due to the high series-field impedance. The result is low running torque. To reduce the field reactance to a minimum, AC series motors are built with as few turns as possible. Armature reaction is overcome by using *compensating windings* (see Figure 8.8) in the pole pieces. As with DC series motors, in an AC series motor the speed increases to a high value with a decrease in load. The torque is high for high armature currents so

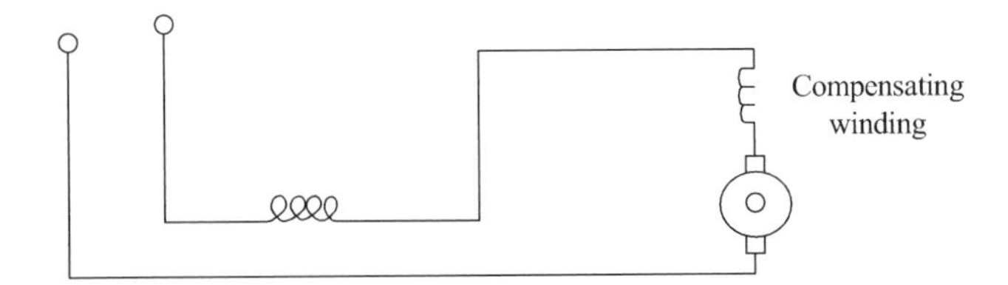

FIGURE 8.8 AC series motor.

that the motor has a good starting torque. AC series motors operate more efficiently at low frequencies. Fractional horsepower AC series motors are called *universal motors*. They do not have compensating windings. They are used extensively to operate fans and portable tools, such as drills, grinders, and saws.

Internal Permanent Magnet (IPM) Motors

Notwithstanding the internal Permanent Magnet (IPM) motors being relatively expensive it is their high power density and maintenance of high efficacy over a high percentage of their operating range that makes the use of IPMs currently the motor of choice of a few EV manufacturers. The high costs are a result of the high prices of magnets and rotor fabrication—both these factors are expensive. Currently, almost all hybrid and plug-in electric vehicles-on-wheels use rare earth permanent magnets and this is a significant part of the expense. The rare earth magnets are expensive due to their limited availability. Even though the cost of rare earth magnets used in IPM motors appears to be increasing (as with almost all other products or services), EV manufacturers are likely to use these magnets until something more practical and less expensive is found as replacements.

Switched Reluctance Motors (SRM)

Because of its simple and rugged construction the switched reluctance motor (SRM) has been gaining interest in industrial applications such as renewable energy systems like wind power and in electric vehicles-on-wheels because it offers a lower cost option that is relatively easy to manufacture. Their rugged structure tolerates high temperatures and speeds. Well, with the good comes the not so good: these motors produce more noise and vibration than comparable motor designs, which is a major challenge for use in vehicles-on-wheels. Another issue is that switched reluctance motors are less efficient than other motor types and require additional sensors and complex motor controllers that increase the overall cost of the electric drive systems.

9 DC/DC Converters

INTRODUCTION

Figure 9.1 shows and highlights DC/DC converters (aka buck-boost converters) used in the electric drive system for many types of electric vehicles-on-wheels. DC/DC converters are used to increase (boost) or decrease (buck) battery voltages (typically 200–450 V) to accommodate the voltage needs of motors and other vehicles and components. If the vehicle electric motor requires higher voltage, such as an internal permanent magnet motor, it will require a boost DC/DC converter. If a component requires lower voltage, such as most vehicle systems (lighting, entertainment, and other accessories), it will require a buck DC/DC converter that reduces the voltage to the 12–42 V level. Research on developing improved converters is active and ongoing. The goal is to increase efficiency, miniaturize or reduce part count, and enable modular (i.e., modular in that it or they contain several parts that serve small functions but combine to serve purpose of the devices and modules that can be removed and replaced), scalable devices (i.e., scalable in the sense that it allows for the capacity to accommodate a greater amount of usage).

TERMINOLOGY AND DEFINITIONS

Step-down (buck converter)—a converter where the output voltage is lower than the input voltage (such as a buck converter).

Step-up (boost converter)—a converter that outputs a voltage higher than the input voltage (such as a boost converter).

Coil-integrated DC/DC converters—decreases mounting space with a small number of components (power control IC, coil, capacity, and resistor) in a single integration solution.

Continuous current mode (CCM)—current and the magnetic field in the inductive energy storage never reach zero—the current fluctuates but never goes down to zero.

Discontinuous current mode (DCM)—current and the magnetic field in the inductive energy storage may reach or cross zero—the current fluctuates during the cycle, going down to zero at or before the end of each cycle.

Galvanic isolation—this is a technique that separates electrical circuits to eliminate stray currents; no direct conduction path is permitted.

Hard switched—in topology refers to transistors switching quickly while being exposed to both full voltage and full current.

Input noise—if the converter loads the input with sharp load edges, the converter can emit RF noise from the supplying power lines. This is usually prevented by using proper filtering in the input stage of the converter.

DOI: 10.1201/9781003332992-9

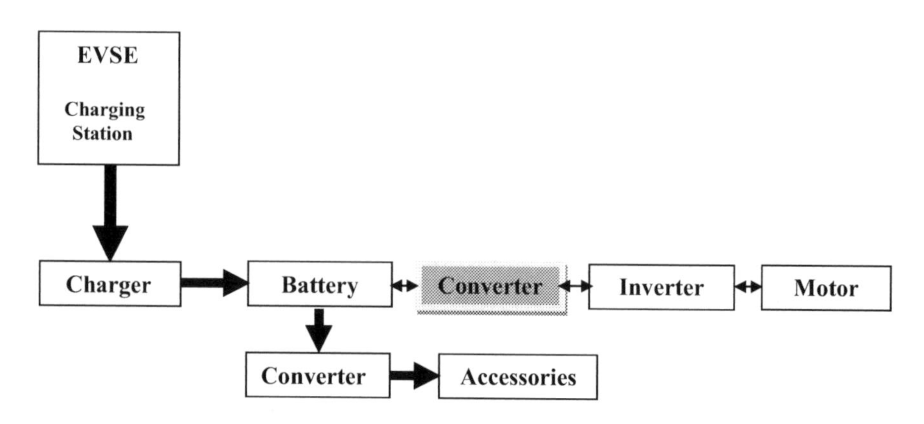

FIGURE 9.1 Vehicle-on-wheels electric drive system components highlighting converter.

Mainstream converter topologies—non-isolated types include boost, buck-boost, and buck; isolated types include flyback, forward, push-pull, half-bridge, and full bridge. Other topologies with numerous variations include SEPIC (single-ended primary-inductor converter), Cuk converter, Current Fed Buck, Tapped Inductors, Multiple Outputs, and Interleaving.

Noise—unwanted electrical and electromagnetic signal noise, typically switching artifacts.

Output noise—even though an ideal DC-to-DC converter produces an output that is flat, with constant output voltage, these converters produce DC output upon which is superimposed some level of electrical noise. Note that switching converters produce switching noise at the switching frequency and its harmonics. Moreover, all electronic circuits have some thermal noise (i.e., noise made by thermal agitation of the electrons flowing in the circuit). Because some electrical/electronic radio-frequency and analog circuits require a power supply with little noise, the use of linear regulators works to maintain a constant voltage output and basically acts like a variable resistor, continuously adjusting to maintain an average value of output results.

Resonant—an LC circuit that shapes the voltage across the transistor and current through it so that the transistor switches when either the voltage or current is zero.

RF noise—switching converters inherently emit radio waves at the switching frequency and its harmonics. Note that switching converters that produce triangular switching currents, such as the Split-Pi, forward converter, or Cuk converter in continuous current mode, produce less harmonic noise than other switching converters (Hoskins, 1997). RF noise causes electromagnet interference (EMI). Acceptable levels depend upon requirements, e.g., proximity to RF circuitry needs more suppression than simply meeting regulations.

DC/DC converters can be classified as isolated or non-isolated converters.

Isolated or non-isolated converters?

Yes.

What is the difference?

Okay, good question. The difference is that the non-isolated DC/DC converter has a single switch and a single diode and also may have inductors and capacitors to store energy and are typically of the boost, buck-boost, or buck converters. Isolated DC/DC converters are derived from the basic topologies defined earlier and also include flyback, forward, push-pull, half-bridge, and full bridge topologies.

Non-isolated and Isolated Converter Topologies (aka Configurations)

Non-isolated Converters

- Boost converter (non-isolated)—is a step-up converter that raises voltage while lowering current from its supply to its load.
- **Buck-boost converter**—is a type of switched-mode power supply that combines the principles of the buck converter and boost converter in a single circuit and provides a regulated DC voltage from either an AC or a DC input.
- **Buck converter** (aka step-down converter)—this DC-to-DC power converter steps down voltage (while drawing less average current) from its supply (input) to its load (output). As a switched-mode-power-supply (SMPS)—often used to convert voltage in computers—it typically contains at least two semiconductors (a diode and a transistor), at present the trend is to replace the diode with a second transistor and at least one energy storage element, a capacitor, inductor, or a combination of both.

Isolated Converters

- Flyback converter—is used in both AC/DC and DC/DC conversion with what is known as *galvanic isolation* between the input and any outputs. This is a buck-boost converter with the inductor split to form a transformer; this facilitates multiplied voltage ratios giving an additional advantage of isolation.
- **Forward converter**—this DC/DC converter uses a transformer to increase or decrease the output voltage based on the transformer's ratio and provides galvanic isolation for the load. This type of converter has multiple output windings, making it possible to provide both higher and lower voltage outputs simultaneously.
- **Push-pull converter**—this converter is a type of DC/DC converter, basically a switching converter that uses a transformer to change the voltage of a DC power supply.
- **Half-bridge converter**—like flyback and forward converters the half-bridge converter is a type of converter that can supply an output voltage either higher or lower than the input voltage and provide electrical isolation via a transformer.
- **Full bridge converter**—this is a DC-to-DC converter configuration that employs four active switching components in a bridge configuration across a power transformer. A full bridge converter, besides reversing the polarity and providing multiple output voltages simultaneously, also is one of the commonly used configurations that offer isolation in addition to stepping up or down the input voltage.

REFERENCE

Hoskins, K. (1997). Making-5V Quiet, section of Linear technology Application Note 84. Accessed 6/12/22 @ http://www.linear.com/docs/4173.

10 Inverters

THE 411 ON ELECTRIC VEHICLES INVERTERS

In Figure 10.1, an inverter is shown in the electric vehicle-on-wheels power train configuration. Not only is the inverter included in the power train of electric vehicles-on-wheels but it also is a principal component. The inverter shown in Figure 10.1 is needed to convert DC energy from a battery (remember batteries produce DC voltage and current) to AC power to drive the motor. An inverter also acts as a motor controller and as a filter to isolate the battery from potential damage from stray currents. In addition, because batteries produce DC voltage and current and most consumer products work on AC, including some electric vehicles-on-wheels, a vehicle power invertor is necessary to use the AC devices on the road.

Note that there is nothing new about using inverters in vehicles-on-wheels and other types of vehicles. At the present time inverters are used in firefighting platforms and ambulances, recreational vehicles, buses, and in non-vehicles-on-wheels such as boats. With the increase in production of electrified vehicles-on-wheels new and potentially stronger demand for inverters is currently occurring—with significant increased use for this purpose in the future.

Conventional vehicle inverters operate off of low voltage and are limited to low power (limits of battery and alternation supply) at 12–24 V DC (VDC) and less than 6 kilowatts (kW). What we call "electric vehicles-on-wheels" in this text (and maybe elsewhere), battery electric, hybrid electric, and other fuel cell applications operate at higher DC (keep this in mind) voltage ranging from 48 to 800 VDC; in the kilowatt range, these vehicles put out 10–200 kW.

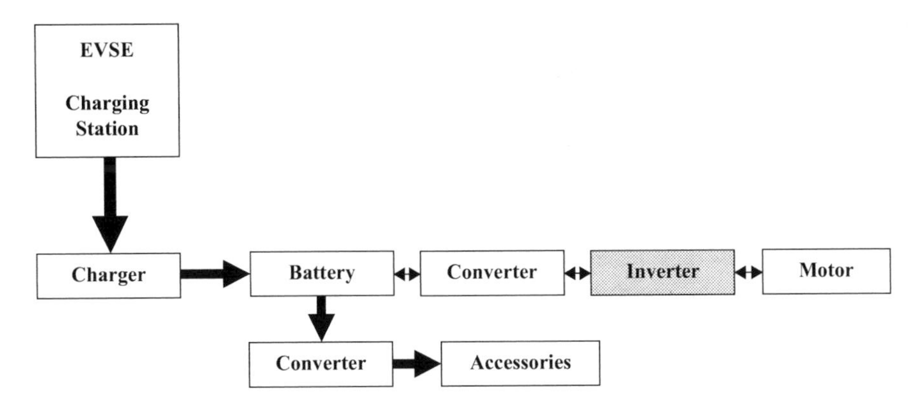

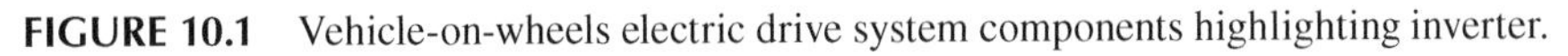

FIGURE 10.1 Vehicle-on-wheels electric drive system components highlighting inverter.

DOI: 10.1201/9781003332992-10

Electric Vehicles—Present and Future

Currently the push is on to switch from hydrocarbon vehicles-on-wheels to electric power-driven vehicles. There are numerous motives behind this movement with global climate change the principal causal-trend factor. The truth be told there is a movement toward electric vehicles-on-wheels that is presently gaining headway in the vehicle-on-wheels market. With gas prices averaging over $5.00/gallon (and seemingly rising) it is not surprising that electric vehicles-on-wheels have become the focus for possibly their desire to purchase one of these vehicles. Gas prices and gas availability have at least made some potential vehicles-on-wheels customers take a look at the electrical vehicles as a possible substitute for hydro-driven powered vehicles.

So, you might think that the manufacturers of electric vehicles-on-wheels rely solely on the high cost of gasoline and diesel as the only selling point needed for their product. This is not the case, however. Competition between manufacturers of vehicles-on-wheels is a significant driver that requires manufacturers and sellers to provide "incentives" or "extras" to attract the buyer's attention and possibly earn a final sale.

Okay, what are the incentives, the extras that are magnets of attention for potential buyers of electric vehicles-on-wheels?

For some vehicles, especially for the on-road vehicles, it is the potential to have included in their vehicle 110 VAC outlets available for standard electrical applications. In addition, the special features of hybrid electric "contractor" trucks are attractive for those who need specialized trucks for their work. A hybrid electric contractor truck has a service body with a ladder rack. Important to a contractor is to have a quality vehicle to transport equipment and assist in completing jobs. Using this type of truck gives the contractor the ability and versality to haul almost anything. A hybrid electric contractor truck comes in various body lengths with varying features and is basically equipped with the perfect body on the job site. Tools and accessories are securely stored and easily accessible. In the case of off-road electric hybrids and EVs the military, construction, and agriculture hybrids and EVs are a real selling feature, this is especially the case because these hybrids and EVs allow for operation away from readily available power and this is a very strong selling feature.

So, it looks like the future of hybrids and EVs and trucks has a glowing outlook. This bright future also extends to the so-called "connected vehicles" (cars). Additionally, high-capacity inverters are a big attraction and sell to the military, agriculture, and construction. Also, fuel cells of various sorts are likely to be innovative technologies that will impact transportation needs.

Note: Fuel cells and various innovations are addressed later in this book.

Well, all of this to this point is positive and encouraging in the transfer from hydrocarbon-powered vehicles to EVs and fuel cell-powered vehicles-on-wheels. With regard to vehicle electrical and mechanical characteristics it can be said that electric vehicles generally have low sensitivity to perturbation. EVs, in most cases, are stand-alone vehicles-on-wheels. With regard to EV dimensions their size and weight is important and should be small and lightweight. Reliability and component life requirements are important and the goal should be to ensure 10-year, 150,000 mile, 5,000 hours vehicle component life.

The bottom line: cost is the overriding figure of merit.

There is more to consider too. Today's EV inverters are missing, lacking, or short of low cost—cost is a huge factor. Also, for a given output, the inverters need to be smaller in size. They also need to be constructed with built-in ability to handle vibration, shock, and extreme temperatures. Inverters must be designed to provide higher power capability and unfortunately at the present time there are few inverter types available for this burgeoning market.

There are important issues related to inverter usage and availability. For example, performance is an issue. High output is required but in a small package, able to survive and function properly in difficult environments. Also, the functionality built into some existing inverter produces is often not required.

And then there is the cost. The truth be told electric vehicles-on-wheels, hybrid electric, and fuel cell vehicles are too expensive for the average person. However, with regard to market share, the market for advanced vehicle inverts is wide open (basically unprotected) to competitors. However, many of the competitors have been tired of waiting for the market to develop and have dropped out, at least for the time being.

11 EV's Carburetor

INTRODUCTION

With regard to operating a battery-powered vehicle-on-wheels most operators simply enter the vehicle, start it up, and drive to wherever headed without giving much thought about the powertrain involved with powering his or her vehicle. Moreover, if the battery-powered vehicle-on-wheels operator is asked about his or her vehicle and specifically if they can explain how the vehicles works, that is, how it operates, they are likely to say, "Well, you get in, you start the motor and you head on out."

There probably is little doubt that this reply is standard. However, if the same operators are asked if they know in greater detail how their battery-operated vehicle-on-wheels operates mechanically or electrically, and so forth, the typical response (based on my unscientific survey of more than 100 participants at a conference I spoke at) was, "Well, you get in, you start that motor and the battery power drives the motor and vehicle and off you go."

I found this to be a standard reply. I asked several of these same individuals if they had any idea about what other electrical or mechanical components made up the powertrain between the battery and traction motor. Not surprisingly only a few of the respondents could name and identify the other components that make up the powertrain, especially the electrical components, the vehicle. However, a few responders mentioned that they understood that some type of electronic device included, what they thought was the computer, the so-called "black box," that ran the entire battery-powered vehicle-on-wheels system.

I did have one respondent who owned an automotive dealership and she said that she sells EVs and hybrids and told me that the EV traction motor is the most essential part of any electric vehicle and she added, "the rest of the components are the 'bells and whistles' such as battery, electronic components such as resistors, capacitors, inductors, diodes, and transistors are included … my mechanics have told me that those electrical/electronic components are essential in providing control of the EV."

This statement about the powertrain components of an EV is the best and most accurate statement I received in describing some of the EV drive train elements.

This chapter concentrates on the EV's controllers and points out that an EV contains more than a battery wired to a motor and that is it—that is that. Truth be told, a battery supplies DC power; with the exception of a DC-brushed motor that will run at a single speed when directly wired to the battery, none of the other motors described earlier in this book will work when wired directly to the battery. In order to use any motors to drive the EV, we must modulate, that is, modify, change the features or characteristics of the voltage coming from the battery. This change in voltage characteristics is either by converting the DC electricity from the battery to AC, or to change the frequency of the AC. These various battery voltage characteristic changes

DOI: 10.1201/9781003332992-11

are necessary to not only operate the vehicle but also to *control* the vehicle operation. The key word just mentioned is "control." A means of controlling speed and other vehicle functions is provided by what are called controllers—a controller allows the generation of various voltages, which in turn provides the ability to control speed and acceleration very much like a carburetor does much like a gasoline-powered vehicle—it makes little sense and is absolutely unsafe to operate a vehicle at one speed only.

CONTROLLERS

The controller in an electric vehicle (EV) on-wheels is commonly called an electronic tool or package; it is the tool/package that operates between the batteries and the motor to control the electric vehicle's speed and acceleration and as mentioned it functions much like a carburetor does in the gasoline-powered vehicle. In the simplest terms what the controller does is transform the battery's direct current (DC) for alternating current (note: for AC motors only) and regulates the energy flow from the battery. The controller, unlike the carburetor, also functions to reverse motor rotation so that the vehicle can go in the reverse direction; it also converts the motor to a generator whereby the kinetic energy produced by motion can be used to recharge the battery when the brake is applied. Note that this technique is known as regenerative braking and is covered later in this text.

In the earlier electric vehicles-on-wheels models with DC motors, a simple variable-resistor-type controlled the acceleration and speed of the vehicle. The problem with this early setup was that full current and power (not voltage, remember voltage is the pump and does not move) was drawn from the battery all of the time. In operation at slow speeds, when full power was not needed, a high resistance was used to reduce the current to the motor. With this particular system, a large percentage of the energy from the battery was wasted as an energy loss in the resistor. The available power was only used at high speeds.

At the present time the modern controllers in use adjust speed and acceleration by an electronic process called pulse width modulation (see Figure 11.1). This modulation is provided by electronic switching devices—because these electronic devices have no physical moving parts and no physical contact they are generally called solid-state switches. There are different types of solid-state switches available for use in electric vehicle traction motor controllers; some of these solid-state switches include transistors, SCRs, MOSFETs, TRIACs, and IGBTs.

We will get back to transistors and electronic switching devices but first we need to describe and discuss the circuit elements that make up the "carburetor" that drives and accelerates the electric vehicle-on-wheels traction motor.

Controller Components/Elements

The rapid rise in producing and using electric vehicles-on-wheels presents new challenges to produce and source and use cutting-edge components that can handle higher temperatures and voltages without sacrificing reliability, availability, and footprint.

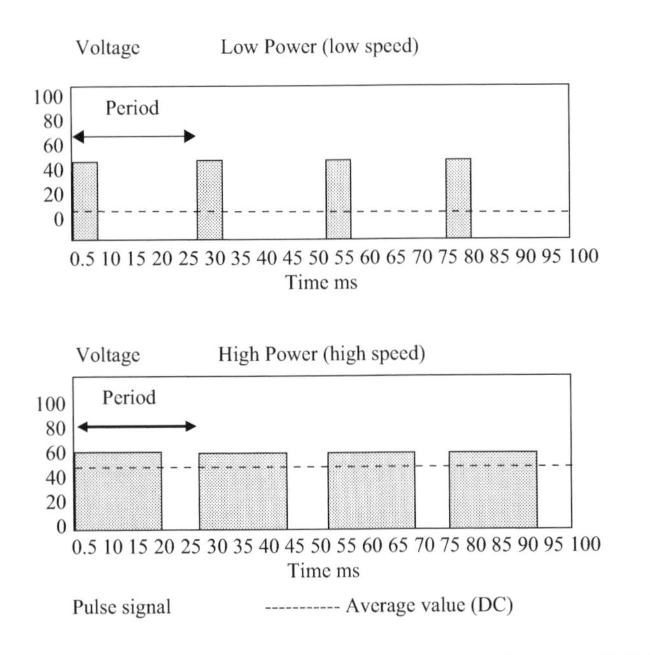

FIGURE 11.1 DC pulse width modulation. (Idaho National Laboratory (INL) Advanced Vehicle Testing Activity accessed June 21 2022 @ https:avt.inl.gov.)

One of the most important cutting-edge components is the capacitor, which is used in various applications in automotive EV electronics.

In order to understand exactly what an electric vehicle-on-wheels controller is and how it works we need to start by reviewing a few of the electrical/electronic elements introduced and discussed earlier in the text and to also discuss diodes and transistors and their functions. These components or elements include resistors, inductors, capacitors, diodes, and transistors. We begin by reviewing transistors and their operation.

Resistors

Recall that resistors have the symbol R, and their resistance is measured in ohms (Ω). Electricity travels through a conductor (wire) easily and efficiently, with almost no other energy released as it passes. On the other hand, electricity cannot travel through a resistor easily. When electricity is forced through a resistor, often the energy in the electricity is changed into another form of energy, such as light or heat. The reason a light bulb glows is that electricity is forced through the tungsten filament, which is a resistor.

Resistors are commonly used for controlling the current flowing in a circuit. A *fixed resistor* provides a constant amount of resistance in a circuit. A *variable resistor* (also called a potentiometer) can be adjusted to provide different amounts of resistance, such as in a dimmer switch for lighting systems. A resistor also acts as a load in a circuit, in that there is always a voltage drop across it. Figure 11.2 shows some of the basic symbols that are used to designate resistors on schematic diagrams.

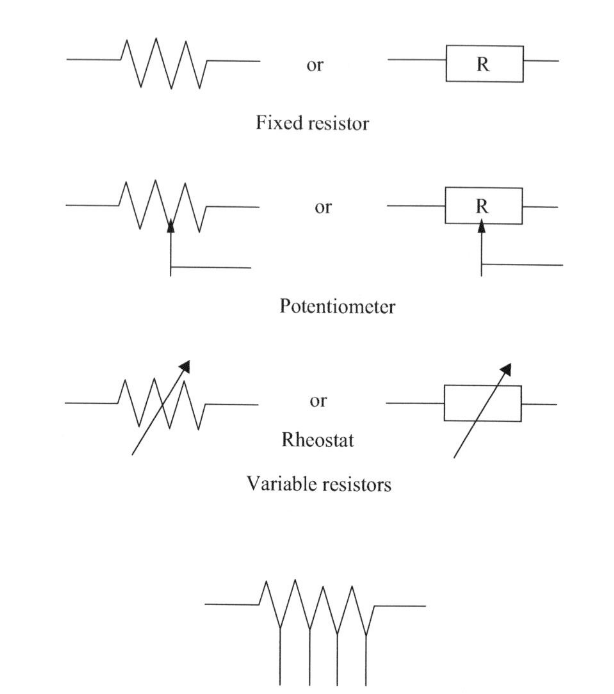

FIGURE 11.2 Resistor symbols.

With regard to resistors used in controllers in electric vehicles-on-wheels their primary purpose is the conversion of electrical energy into rotational motion in an electric traction motor. The earliest electric vehicle controllers used switches and resistors to directly control motor speed. With time and advances in technology other components have replaced direct resistor use in controlling acceleration and speed. Note that although not directly involved in controlling motor speed in modern electric vehicles-on-wheels resistance is always present whenever there is current flow in an electrical circuit. Also, modern controller electronics use and count on resistance in one form or another as a key element in their construction.

EVs use *power resistors*, with a standard power rating of 5 W designed to resist and dissipate a large amount of power (standard resistors do not dissipate huge power). Power resistors are available in a compact physical package and are designed to dissipate high power by keeping their size small if possible.

Inductors

Inductors provide one of the essential electromagnetic principles that allows EV controllers to function. Recall that earlier we explained the following key points about magnetic fields:

- A field of force exists around a wire carrying a current.
- This field has the form of concentric circles around the wire, in planes perpendicular to the wire, and with the wire at the center of the circles.

- The strength of the field depends on the current. Large currents produce large fields; small currents produce small fields.
- When lines of force cut across a conductor, a voltage is induced in the conductor.

Earlier we studied circuits that have been *resistive* (i.e., resistors presented the only opposition to current flow). Two other phenomena—inductance and capacitance—exist in DC circuits to some extent, but they are major players in AC circuits. Both inductance and capacitance present a kind of opposition to current flow that is called *reactance*, which we will cover later. Before we examine reactance, however, we must first study inductance and capacitance.

Okay, recall that *Inductance* is the characteristic of an electrical circuit that makes itself evident by opposing the starting, stopping, or changing of current flow. A simple analogy can be used to explain inductance. We are all familiar with how difficult it is to push a heavy load (a cart full of heavy materials, etc.). It takes more work to start the load moving than it does to keep it moving. This is because the load possesses the property of *inertia*. Inertia is the characteristic of mass that opposes a *change* in velocity. Therefore, inertia can hinder us in some ways and help us in others. Inductance exhibits the same effect on current in an electric circuit as inertia does on velocity of a mechanical object. The effects of inductance are sometimes desirable—sometimes undesirable.

Important Point: Simply put, inductance is the characteristic of an electrical conductor that opposes a change in current flow.

Because inductance is the property of an electric circuit that opposes any *change* in the current through that circuit, if the current increases, a self-induced voltage opposes this change and delays the increase. On the other hand, if the current decreases, a self-induced voltage tends to aid (or prolong) the current flow, delaying the decrease. Thus, current can neither increase nor decrease as fast in an inductive circuit as it can in a purely resistive circuit.

In AC circuits, this effect becomes very important because it affects the *phase* relationships between voltage and current. Earlier we learned that voltages (or currents) can be out of phase if they are induced in separate armatures of an alternator. In that case, the voltage and current generated by each armature were in phase. When inductance is a factor in a circuit, the voltage and current generated by the *same* armature are out of phase. We shall examine these phase relationships later. Our objective now is to understand the nature and effects of inductance in an electric circuit.

The unit for measuring inductance, L, is the *Henry* (named for the American physicist, Joseph Henry), abbreviated h and normally written in lower case, henry. Figure 11.3 shows the schematic symbol for an inductor. An inductor has an inductance of 1 henry if an emf of 1 volt is induced in the inductor when the current through the inductor is changing at the rate of 1 ampere per second. The relation between the induced voltage, inductance, and rate of change of current with respect to time is stated mathematically as

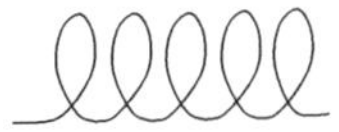

FIGURE 11.3 Schematic symbol for an inductor.

$$E = L\frac{\Delta I}{\Delta t} \tag{11.1}$$

where

E = the induced emf in volts
L = the inductance in henry
ΔI = the change in amperes occurring in Δt seconds

Note: The symbol Δ (Delta) means "a change in …."

The henry is a large unit of inductance and is used with relatively large inductors. The unit employed with small inductors is the millihenry (mh). For still smaller inductors the unit of inductance is the microhenry (μh).

As previously explained, current flow in a conductor always produces a magnetic field surrounding, or linking with, the conductor. When the current changes, the magnetic field changes, and an emf is induced in the conductor. This emf is called a *self-induced emf* because it is induced in the conductor carrying the current.

Note: Even a perfectly straight length of conductor has some inductance.

The direction of the induced emf has a definite relation to the direction in which the field that induces the emf varies. When the current in a circuit is increasing, the flux linking with the circuit is increasing. This flux cuts across the conductor and induces an emf in the conductor in such a direction to oppose the increase in current and flux. This emf is sometimes referred to as **counterelectromotive force** (cemf). The two terms are used synonymously throughout this manual. Likewise, when the current is decreasing, an emf is induced in the opposite direction and opposes the decrease in current.

Important Point: The effects just described are summarized by **Lenz's Law**, which states that the induced emf in any circuit is always in a direction opposed to the effect that produced it.

Shaping a conductor so that the electromagnetic field around each portion of the conductor cuts across some other portion of the same conductor increases the inductance. A loop of conductor is looped so that two portions of the conductor lie adjacent and parallel to one another. These portions are labeled Conductor 1 and Conductor 2. When the switch is closed, electron flow through the conductor establishes a typical concentric field around **all** portions of the conductor. The field is shown in a single plane (for simplicity) that is perpendicular to both conductors. Although the field originates simultaneously in both conductors it is considered as originating in Conductor 1 and its effect on Conductor 2 will be noted. With increasing current, the field expands outward, cutting across a portion of Conductor 2. The resultant induced emf in Conductor 2 is shown by the dashed arrow. Note that it is in **opposition** to the battery current and voltage, according to Lenz's Law.

Four major factors affect the self-inductance of a conductor, or circuit.

1. Number of turns: Inductance depends on the number of wire turns. Wind more turns to increase inductance. Take turns off to decrease the inductance.

2. Spacing between turns: Inductance depends on the spacing between turns, or the inductor's length.
3. Coil diameter: The larger-diameter inductor has more inductance.
4. Type of core material: **Permeability**, as pointed out earlier, is a measure of how easily a magnetic field goes through a material. Permeability also tells us how much stronger the magnetic field will be with the material inside the coil. The inductance of a coil is affected by the magnitude of current when the core is a magnetic material. When the core is air, the inductance is independent of the current.

Key Point: The inductance of a coil increases very rapidly as the number of turns is increased. It also increases as the coil is made shorter, the cross-sectional area is made larger, or the permeability of the core is increased.

Unlike the fundamental relationship for a resistor, which states that the voltage is proportional to the current, in the case of an inductor, the voltage is proportional to the derivative of the current. As we will see later, the quality of inductors is one of the essential electromagnetic principles that allow controllers to function.

Capacitors

Earlier, we learned that *capacitance* is the property of an electric circuit that opposes any change of *voltage* in a circuit. That is, if applied voltage is increased, capacitance opposes the change and delays the voltage increase across the circuit. If applied voltage is decreased, capacitance tends to maintain the higher original voltage across the circuit, thus delaying the decrease.

Recall that capacitance is also defined as that property of a circuit that enables energy to be stored in an electric field. Natural capacitance exists in many electric circuits. However, in this manual, we are concerned only with the capacitance that is designed into the circuit by means of devices called capacitors.

As mentioned earlier a *capacitor*, or condenser, is a manufactured electrical device that consists of two conducting plates of metal separated by an insulating material called a *dielectric* and the schematic symbols for popularly used capacitors is shown in Figure 11.4. (Note: the prefix "di–" means "through" or "across").

When a capacitor is connected to a voltage source, there is a short current pulse. A capacitor stores this electric charge in the dielectric (it can be charged and discharged, as we shall see later). To form a capacitor of any appreciable value, however,

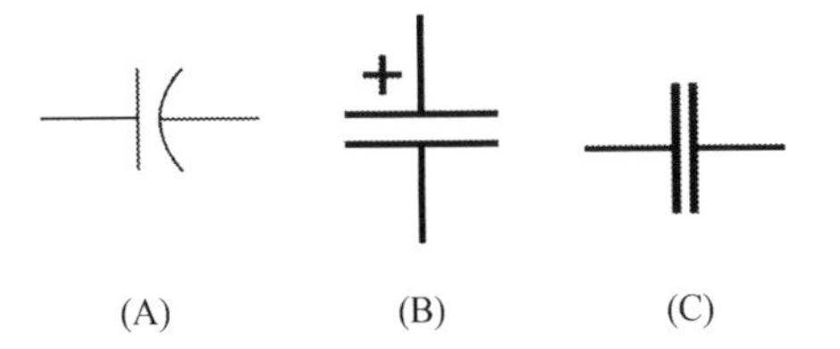

FIGURE 11.4 Schematic symbols for capacitors: A is symbol for standard electronic capacitor; B is symbol for electrolytic (polarity type) capacitor; and C is for a ceramic capacitor.

the area of the metal pieces must be quite large and the thickness of the dielectric must be quite small.

Key Point: A capacitor is essentially a device that stores electrical energy.

The capacitor is used in a number of ways in electrical circuits. It may block DC portions of a circuit since it is effectively a barrier to direct current (but not to AC current). It may be part of a tuned circuit—one such application is in the tuning of a radio to a particular station. It may be used to filter AC out of a DC circuit. Most of these are advanced applications that are beyond the scope of this presentation; however, a basic understanding of capacitance is necessary to the fundamentals of AC theory.

Important Point: A capacitor does not conduct DC current. The insulation between the capacitor plates blocks the flow of electrons. We learned earlier there is a short current pulse when we first connect the capacitor to a voltage source. The capacitor quickly charges to the supply voltage, and then the current stops.

Important Point: In a capacitor, electrons cannot flow through the dielectric, because it is an insulator. Because it takes a definite quantity of electrons to charge ("fill up") a capacitor, it is said to have **capacity**. This characteristic is referred to as *capacitance*.

Capacitors used for EVs are multilayer ceramic capacitors (MLCCs) because they are the perfect component for EV electronics and subsystems.

Perfect component for use in EVs?

Yes, and this is the case because MLCCs are tiny and have high temperature and offer an easy surface-mount form factor.

What exactly are the capacitor requirements for EVs?

Good question. And because MLCCs have become one of the key elements for control electronics in electric vehicles-on-wheels they must meet tough demands such as those listed as follows.

- **Increased voltage ranges**—EV systems are based on high-voltage battery systems, as a result capacitors must be rated for increased voltage ranges such as 250–400 V for plug-in hybrid electrical vehicles-on-wheels, 800 V for commercial vehicles, and 48 V for hybrid electric vehicles (HEVs).
- **Ultra-fast charging stations**—the only way EVs can compete with hydrocarbon-based vehicles that can refill in a few minutes, EVs need to recharge fast.
- **Able to withstand high temperatures**—a high-temperature environment can be detrimental to component operation. The drivers behind high-temperature environments in EVs are the combination of increased voltage, higher convert frequencies, and small footprint (miniaturization) combined to create the high-temperature environment in the EV control system.
- **Harsh operating conditions**—automotive-grade capacitors need to meet rigorous standards and withstand the most challenging environments that require a cut above mechanical performance under extremely harsh operating conditions.

MCCL Use in EVs

In Figure 11.5 MCCL use as main subsystems in EVs and PHEVs is illustrated. Their use in these five common subsystems is important to the operation of all varieties of electric vehicles-on-wheels. The truth be told the advancements in technology involved in EVs, HEVs, and PHEVs have sparked a transformation in capacitor technology (still evolving) targeting subsystem electronics. Because of the higher temperatures inside the control circuits, conversion plastic film capacitors are no longer suitable for all applications, and ceramic MLCCs are now increasingly used in EV subsystems (see Figure 11.5).

Now, let's take a look at, and examine the challenges and requirements of these subsystems as well as the types of MLCCs normally used in each one.

Okay, from Figure 11.5 one of the subsystems in electric vehicle-on-wheels is related to battery management. Simply, the battery management system (BMS) is designed to control and monitor the cells within a battery stack. So, what does this include? Well, it includes protecting the vehicle's battery from under/over-charging. Moreover, the battery management system performs cell balancing by managing the current among cells and also monitors battery cell's voltage, current, and temperature and basically the batteries' overall condition. Note that the battery management systems also work to collect and provide diagnostic data for the battery.

Another subsystem shown in Figure 11.5 is the on-board charger (OBC). The traction motor battery is charged, usually in the range of 48–800 V by the on-board charger. The OBC functions to convert AC from the electric grid into DC that charges the vehicle's battery. It also provides power factor correction (PFC) to shape the input current to a power supply, maximizing efficiency and reducing harmonics.

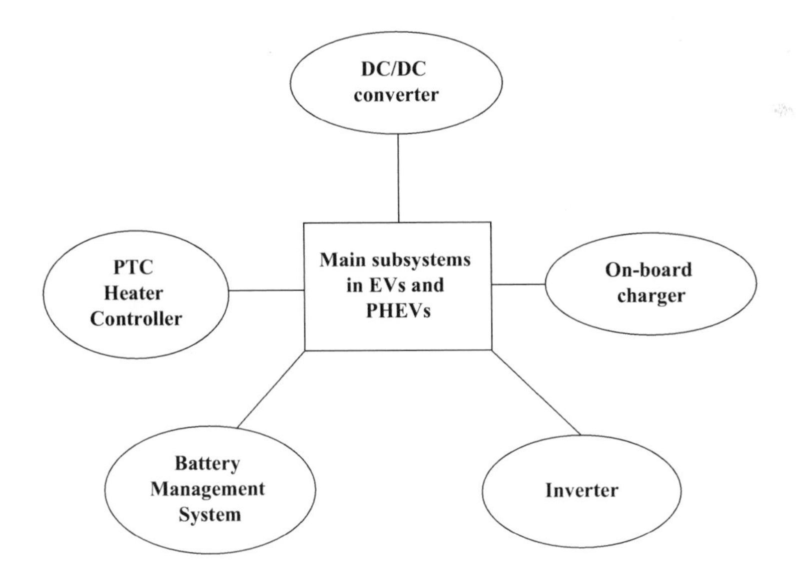

FIGURE 11.5 The main subsystems in EVs and PHEVs where MCCLs are commonly used.

Power factor is important.

Why?

Power factor is important because electric vehicles-on-wheels and the grid whose loads have low power factor require greater generation capacity than what is actually metered. If connected to the grid, this imposes a cost on the electric utility that is not otherwise recorded by the energy and demand charges. There are two types of power that make up the total or apparent power supplied by the electric utility. The first is *active* (also called *true* power) power. Measured in kilowatts, it is the power used by the equipment to produce work. The second is the *reactive* power. This is the power used to create the magnetic field necessary for induction devices to operate. It is measured in kVARs (kilo-volt-amp-reactive).

- **Active power (true power) (P)**—power that performs work measured in Watts (W)
- **Reactive power**—power that does not perform work (sometimes called "wattless power") measured in VA reactive (VAr)
- **Complex power**—the vector sum of the true and reactive power measured in volt amps (VA)
- **Apparent power**—the magnitude of the complex power measured in volt amps (VA)

The vector sum of the active and reactive power is called complex power. Power factor is the ratio of the *active* power to the *apparent* power. The power factor of fully loaded induction motors ranges from 80% to 90% depending on the type of motor and the motor's speed. Power factor deteriorates as the load on the motor decreases. Other electrical devices such as space heaters and old fluorescent or high discharge lamps also have poor power factor. Power factor may be leading or lagging. Voltage and current waveforms are in phase in a resistive AC circuit. However, reactive loads, such as induction motors, store energy in their magnetic fields. When this energy gets released back to the circuit it pushes the current and voltage waveforms out of phase. The current waveform then lags behind the voltage waveform.

Improving power factor is beneficial as it improves voltage, decreases system losses, frees capacity to the system, and decreases power costs where fees for poor power factor are billed. Power factor can be improved by reducing the reactive power component of the circuit. Adding capacitors to an induction motor is perhaps the most cost-effective means to correct power factor as they provide reactive power. Synchronous motors are an alternative to capacitors for power factor correction.

In addition to power factor adjustment the on-board charger works to adjust the produced DC voltage up and down to provide the correct DC levels to the battery. Ceramic capacitors (because they can handle the heat) are included in the primary section of the OBC and may function as an EMI or AC line filter. The PFC input capacity works to smooth the pulsating DC voltages produced from the AC rectifier and must have comparably high capacitance. When included in the package capacitor works as an output filter to remove the ripple copoint of the AC current and smooth the output voltage of the power converter.

With regard to the DC-DC converter shown in Figure 11.5, this component as used in EV subsystems is required to transfer energy between the high-voltage battery and the 12 V low-voltage systems. The high-voltage system provides the power needed for the traction motor and accessories such as the starter and air-conditioning. The low-voltage system is used for components such as sensors, safety and infotainment systems (i.e., heating, AC, radio, phone, GPS, cabin lighting, and internet access). Again because of high-temperature conditions ceramic capacitors instead of film capacitors are used in the DC-DC converter.

Another electric vehicle-on-wheels subsystem commonly using MCCLs is the inverter (see Figure 11.5). Recall that an inverter is needed to convert DC energy from a battery (remember batteries produce DC voltage and current) to AC power to drive the motor. An inverter also acts as a motor controller and as a filter to isolate the battery from potential damage from stray currents. In addition, because batteries produce DC voltage and current and most consumer products work on AC, including some electric vehicles-on-wheels, a vehicle power inverter is necessary to use the AC devices on the road. Note that the inverter converts DC to AC during acceleration and AC to DC during braking.

Another important subsystem in electric vehicles-on-wheels shown in Figure 11.5 is the positive-temperature-coefficient (PTC) heater controller (aka self-regulating heater). The PTC is an electrical resistance heater whose resistance increases significantly with temperature (a large positive temperature coefficient of resistance producing a large amount of heat when tis temperature is low). The PTC is required in EVs because unlike gas/diesel-powered vehicles where the combustion engine's waste heat, the EVs do not have that kind of waste heat available for use in their HVAC systems. The PTC provides a comfortable cabin environment for passengers and also provides optimal operating temperatures for the batteries. Note that this allows the batteries to properly start and charge the vehicle in cold temperatures. Some of these PTCs are designed to have a sharp change in resistance at a particular temperature. These components are self-regulating because they tend to maintain the temperature, even if the applied voltage or heat load changes (Process Heating, 2005; Fabian, 1996). Note that PTC heaters use specialized materials that when cold have lower electrical resistance, enabling full current to flow and heat to be generated. The important thing is that this property acts as an automatic safety feature that limits the current flow and prevents overheating. This subsystem requires high-voltage capacitors for snubber (i.e., a device used to "snub" electrical impulses) and primary-secondary isolation purposes.

Supercapacitors: The Need

The worldwide market is bursting in mainstream adoption and in demand for electric vehicles-on-wheels as the focus is pushed toward fuel economy improvements. Then there is the American way of looking for a way and finding a way in which to make a dollar. Currently, because the ongoing search to find ways in which to make a dollar begins with finding a better way to store energy and to make energy more quickly available when needed beyond the performance and capacity of today's lithium batteries, the terms ceramic capacitors, supercapacitance, double-layer capacitance, and pseudocapacitance have come to the forefront of research and discussion. For the record supercapacitors are capacitors that possess high capacity with a capacitance value much higher than other capacitors, but with lower voltage limits.

A pseudocapacitor is part of an electrochemical capacitor, and forms together with an electric double-layer capacitor to create a supercapacitor. The reason discussion has increased about supercapacitors—by any name—is that researchers are actively studying ways in which to improve energy storage methods, capacity, and devices. This is important in electric vehicle science because when electricity produced by batteries is used to power electric vehicles-on-wheels and for other ancillary purposes devising a method of increasing storage capability and providing other vehicle services via supercapacitors is intriguing and can lead to a beneficial outcome.

The bottom line: supercapacitors have a well-investigated run in their use for battery combinations in EVs and HEVs. The ability of supercapacitors outpacing batteries with regard to charging time, their stable, very stable, electrical properties, and broader temperature range and longer lifetime are very positive characteristics that continue to be experimented with to increase efficiency, reduce weight and volume that is ongoing and will continue to advance with time and experiment.

Diodes: The One-Way Valves

The schematic symbol for a diode is shown in Figure 11.6. A *diode* is a two-terminal component that conducts current primarily in one direction; it has zero or low resistance in one direction, and usually infinite resistance in the other. In actual practice and usage, you will find it hard to find a diode in electric vehicles-on-wheels that is not also accompanied by an attached heat sink. It is all about heat. Diodes can fail in their breakdown region because they can get too hot and can fail permanently—this is the case because the current and voltage are high in the negative direction.

Transistors

The schematic symbol for a transistor is shown in Figure 11.7. Simply, a transistor is a semiconductor device used either to amplify or switch electrical signals and power. It is an understatement when we assert that the transistor is one of the basic building blocks; it is the modern foundation of modern electronics. The semiconductor material it is made from usually has three terminals (at least) for connection to an electronic circuit. A transistor can amplify a signal. A voltage or current applied to one pair of the transistor's terminal controls the current through another pair of terminals. When this occurs the signal or controlled output power can be higher than the controlling input power and this is when the amplification happens.

The mostly commonly used power transistor is the MOSFET. However, as of 2022 the IGBT (insulated gate bipolar transistor) is the second most commonly used and is a transistor of choice for use in many EVs. The schematic symbol for an IGBT is shown in Figure 11.8.

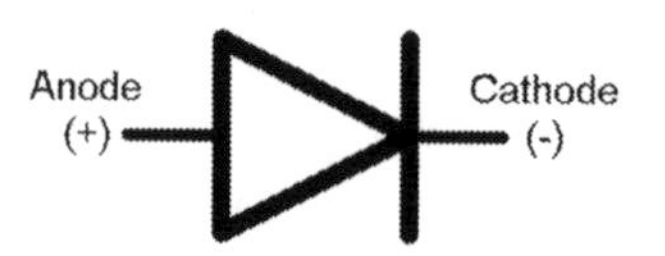

FIGURE 11.6 Schematic symbol for a diode.

FIGURE 11.7 Schematic symbol for a transistor.

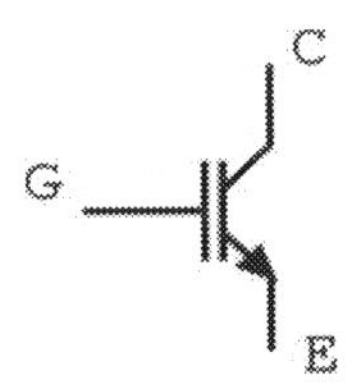

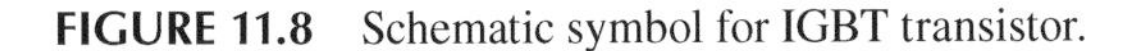

FIGURE 11.8 Schematic symbol for IGBT transistor.

In use the IGBT is primarily used as a high-efficiency and fast electronic switch in electric vehicles-on-wheels. Note that in EV controllers the IGBT is a pulse of varying width, which is applied to the gate. To allow just the right amount of energy to flow from the battery into the remainder of the circuit the width of the pulse is controlled and stored in a capacitor or inductor or both. From these storage device or devices, the energy is smoothly made to rotational motion by the traction motor.

A CHANGE OF STATE

The operation of electric traction motor controllers is based on a change of state. To begin with let's look at standard AC and DC motor controllers used in industrial and/or other applications. The purpose of this type of controller is first to start and stop the motor. Secondly, the standard motor controller provides short circuit protection via a fuse or circuit breaker. Thirdly, the standard motor controller protects the motor with low-voltage/high-voltage protection. Lastly, the standard motor controller basically changes the state of operation of the motor (i.e., for example, increases or decreases the speed of the traction motor).

Changing the state of some action is a common practice in nature as well as in mechanics. For example, in nature and in controlled operations evaporation of a substance like water occurs when the liquid reaches a high enough energy level to change from liquid to the gas state. This natural change also includes condensation melting, freezing, sublimation, and deposition. In the electrical operation of a vehicle-on-wheels the controller performs changes of state so that the traction motor can be controlled and made to respond as directed by the driver. Many control functions occur in the EV but, of course, one of the most important controls is speed.

Although speed control of the traction motor is accomplished by either a DC or AC motor the operation of each is different. In the DC motor to increase speed the controller must increase voltage to the motor terminals—this action increases current flow in the motor windings and in turn increases torque. The controller in the AC motor operates differently. When the controller needs to increase speed, it's all about increasing the frequency waveform; then the controller modulates the frequency waveforms to actually control the vehicle's speed.

Note that the change of state in the DC or AC controllers roughly occurs as just mentioned and is described in more detail in the following sections.

DC Controllers

Without the simple DC step-down controller the modern electric vehicle-on-wheels would not be possible. A number of different DC/DC controllers are available but the underlying function is basically the same. The design chosen can be a step-down control, which turns a high voltage into a low voltage, or a step-up controller, which turns a low-voltage input into a high-voltage output.

Typically, the battery of an electric vehicle-on-wheels has an output that is high—several hundred volts DC. This is fine and dandy, so to speak however because the components inside the vehicle-on-wheels vary in their voltage requirements, with most running on a much lower voltage, including air-conditioning, radio, dashboard readouts, built-in computers, backup cameras, and other displays.

In design of the electric vehicle-on-wheels it is important to include a battery that is not too heavy. This is a challenge because the DC motor uses up to three times the voltage provided by the battery. Therefore, the right controller is needed to reduce battery size and weight. In accomplishing this it is all about Ohm's Law. This is the case, of course, because DC/DC controllers are a subset of electrical engineering and there are many ways in which to accomplish powering of the electric vehicle-on-wheels, but all require compliance with the principles of electricity.

The simple step-up controller works to generate a series of on-off pulses. These pulses must be polished, that is, they must be smoothened into a consistent DC supply whereby the current is constant and the voltage is determined by the change of state brought about by the duration of on states relative to the off cycle. This change of state is accomplished by using a combination of capacitors and inductors.

If the electric vehicle-on-wheels traction motor is powered by AC instead of DC, the controller functions in much the same way as the DC/DC controller, except that it runs the incoming signal through a series of diodes, which clips the negative portion of the signal. The result is a positive wave that can be smoothed into the desired voltage using a combination of resistors, inductors, and capacitors.

THE BOTTOM LINE

The electric vehicle-on-wheels motor controller is the core control device used to control the starting, running, advancing and retreating, speed, stopping of the electric vehicle traction motor, and other electronic components of the electric vehicle. For our purposes just consider the traction motor controller the brain of the electric vehicle-on-wheels.

REFERENCES

Fabian, J. (1996). Heating with PTC thermistors. EBN. *UBM Cannon. 41. 12A.*

Process Heating, 2005. How to specify a PTC Heater for an Oven of Similar appliance. Accessed 06/25/22 @ https://www.world.cast.org/issn/1077.

12 Regenerative Braking and More

SLOWING AND STOPPING

When operating a well-maintained and fully operational electric vehicle-on-wheels and when the driver starts the vehicle and steps on the accelerator the vehicle will go forward or in reverse, depending on the driver's desired selection. When the driver wants to go slow, he or she is able to manipulate the accelerator to attain the desired speed and hopefully the safe speed.

Okay, all that just stated is nothing new about driving a well-maintained and fully operational vehicle-on-wheels. So, the question is what is new about driving a vehicle-on-wheels as compared to driving an electric vehicle-on-wheels? Well, to start with it is safe and correct to say that there are several differences between internal combustion-powered vehicles and the electric vehicle-on-wheels. To this point several of the differences between the two modes of transportation have been pointed out and discussed. However, the slowing and stopping of an internal combustion-powered vehicle and slowing and stopping the electric vehicle-on-wheels is different. They are different in the technology that is used: the electric vehicle-on-wheels braking operation is to slow and to execute a full stop. The difference in the braking system used in electric vehicles-on-wheels is what this chapter is all about.

THE 411 ON REGENERATIVE BRAKING

Regenerative braking is one of the major advantages in electric or hybrid electric vehicles. Why? Well, regenerative braking allows an electric or hybrid electric vehicle to collect electricity as it decelerates. Note that traditional braking results in a large amount of lost energy, which in traffic leads to increased consumption and wear on brakes. In an electric vehicle-on-wheels (EVs), regenerative braking is performed by the electric motor, not by mechanical brakes. This is an advantage because EV drivers can use their brakes less. Note that EVs also have conventional braking systems.

How Does Regenerative Braking Work?

As mentioned, in conventional gas-powered vehicles-on-wheels braking results in a good deal of lost energy. This is not the case in an EV's regenerative system. However, keep in mind that it is not possible to capture all of the energy used in the braking process for other purposes like recharging the battery. The energy captured is that energy that has built up in accelerating the vehicle. When sudden stopping is initiated, braking energy is lost.

DOI: 10.1201/9781003332992-12

Before moving on with a detailed discussion of what regenerative braking is all about and how it works in EVs, it is important to look at some basic science by the means of the math that must be considered in construction of the EV.

One of the first concerns for those involved with regenerative braking is to determine the kinetic energy involved in the system. In those systems that store kinetic energy in their flywheel system, the *kinetic energy recovery systems (KERS)* recovers a moving vehicle's kinetic under braking when a flywheel is to store reservoir fashion for later use under acceleration. When the flywheel is used to store the energy, its energy can be described by the general energy equation:

$$E_{in} - E_{out} = \Delta E_{system} \tag{12.1}$$

where

E_{in} = the energy into the flywheel
E_{out} = the energy out of the flywheel
ΔE_{system} = the change in energy of the flywheel

Okay, when we use this equation we assume that during braking there is no change in the potential energy, enthalpy (that is, the total heat—equal to the energy of the system plus the product of pressure and volume) of the flywheel, pressure or volume of the flywheel—this allows us to consider only kinetic energy. As the vehicle is braking, no energy is dispersed by the flywheel. Basically, the only energy into the flywheel is the initial kinetic energy of the car. Equation (12.1) can be simplified to:

$$\frac{mv^2}{2} = \Delta E_{fly} \tag{12.2}$$

where

m is the mass of the vehicle
v is the initial velocity of the vehicle just before braking

A percentage of the initial kinetic energy of the vehicle is collected by the flywheel—this percentage can be represented by $\eta_{fly.}$ It is the rotational kinetic energy that the flywheel stores. This is all about efficiency and because the energy is kept as kinetic energy and not transformed into any other type of energy. On the amount of energy that the flywheel can store be advised it is limited and depends on the maximum amount of rotational kinetic energy. How is this determined? Well, it's all about inertia of the flywheel and its angular velocity. When the vehicle is sitting idle little rotation kinetic energy is lost over time so the initial amount of energy in the flywheel can be assumed to equal the final amount of energy distributed by the flywheel. Therefore, we can determine that the amount of kinetic energy distributed by the flywheel is:

$$KE_{fly} = \frac{\eta_{fly} mv^2}{2} \tag{12.3}$$

With regard to regenerative braking, it has a similar equation for the mechanical flywheel. Basically a two-step regenerative braking process involves the motor/generator and the battery. It is the generator that initiates the kinetic energy, whereby it is transformed into electrical energy by the generator. Note that this energy is converted to chemical energy by the battery. It is important to point out that this process is less efficient than the flywheel. Having said this the efficiency of the generator can be represented by:

$$\eta_{gen} = \frac{W_{out}}{W_{in}} \tag{12.4}$$

where

W_{in} = the work into the generator
W_{out} = the work produced by the generator

So, let's look at how to determine the power produced by the generator using Equation (12.5).

$$P_{gen} = \frac{\eta_{gen} mv^2}{2\Delta t} \tag{12.5}$$

where

Δt is the amount of time the car brakes
m is the mass of the car
v is the initial velocity of the car just before braking

Efficiency Comparison and Contrast

The U.S. Department of Energy (DoE) provides an interesting comparison and contrast between vehicles-on-wheels powered by internal combustion engines and electric vehicles-on-wheels. With regard to vehicles-on-wheels powered by internal combustion engines (ICE) DoE provides Figure 12.1 showing the efficiency of an ICE vehicle in urban driving compared to the same ICE vehicle during highway driving.

Some of the key points made are:

- The higher the recuperation efficiency, the higher the recuperation.
- The higher the efficiency between the electric motor and the wheels, the higher the recuperation.
- The higher the braking proportion, the higher the recuperation.
- The recuperation rate depends on the driving (where the driving is being done); for example, on motorways, the rate is roughly 3% and in urban areas (cities) the rate can amount to 14%.

The Bottom Line

In electric vehicles-on-wheels, regenerative braking is an energy recovery mechanism. It is a mechanism that slows down a moving vehicle by converting its kinetic energy into a form that can be either used immediately or stored until needed.

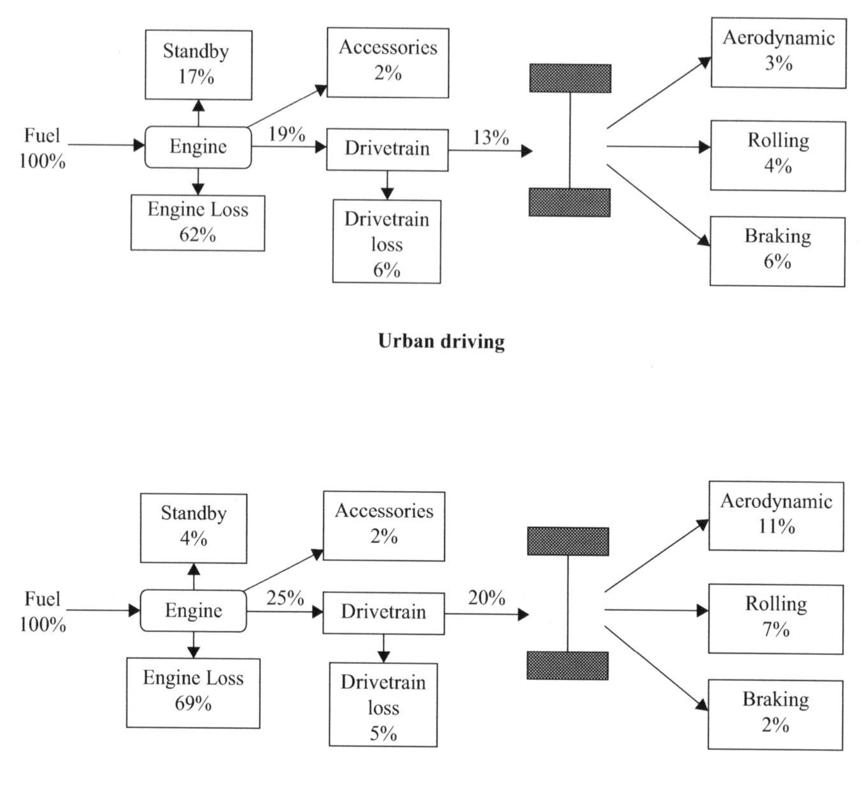

Highway driving

FIGURE 12.1 Efficiency comparison between ICE operation in urban driving and highway driving. (An adaption from US DOE (2022) *All-Electric Vehicles*. Office of Energy Efficiency & Renewable Energy Accessed July 1, 2022 @https://fueleconomy.gov/feg/evtech.shtml.)

Moreover, in this mechanism the electric traction motor uses the vehicle's momentum to recover energy that would otherwise be lost to the brake discs as heat. This differs with other conventional braking systems, where the excess kinetic energy is converted to unwanted waste heat due to friction in the brakes, or with rheostatic brakes (i.e., where DC drive motors are used as generators and convert the kinetic energy of the motor rotor and connected load to electrical energy) where the energy is recovered by using electric motors as generators but is immediately dissipated as heat in resistors. A huge advantage of improving the overall efficiency of the vehicle via regeneration is that the life of the braking system will be extended because the mechanical parts will not wear out very quickly.

13 Battery Power Alternatives

INTRODUCTION

Till this point in the text the focus has been on battery-powered vehicles-on-wheels, and for vehicles powered by electricity this makes sense. However, it is important to point out that there are technologies, systems, and products that are in line with providing motive power for vehicles-on-wheels but they are not battery-powered systems. Specifically we can simply call these alternative technologies, systems, and products as fuel cells. This chapter is all about fuel cells. Why? Well, even though electric-powered vehicles-on-wheels is the present focus of many manufacturers and some others it is important to recognize that technology is dynamic (ongoing) and we are constantly trying to develop power systems for the future of transportation needs and other purposes. So, to be inclusive in regard to alternate ways in which to power vehicles-on-wheels, specifically, it is important to point out that there are other ways to store power and use it to power several of our various systems including vehicles-on-wheels.

FUEL CELLS

> I believe fuel cell vehicles will finally end the hundred-year reign of the internal combustion engine as the dominant source of power for personal transportation.
>
> It's going to be a winning situation all the way around—consumers will get an efficient power source, communities will get zero emissions, and automakers will get another major business opportunity—a growth opportunity.
>
> **—William C. Ford, Jr., Ford chairman, International Auto Show, January 2000**

> I believe that water will one day be employed as a fuel, that hydrogen and oxygen which constitute it, used singly or together, will furnish an inexhaustible source of heat and light.
>
> **—Jules Verne, *Mysterious Island*, 1874**

With regard to the term "cell," depending on education level, social, cultural, and/or economic background, it may conjure up initial meanings that are as diverse in variety as are the colors, sizes, and shapes of lightning bolts (by the way, lightning bolts are a huge source of untapped "renewable" energy). Some may think of the term cell as referring in terms related to plant cell, animal cell, cell structure, cell biology, cell diagram, cell membrane, human memory, cell theory, cell wall, cell parts, cell function, honeycomb cell, prison cell, electrolytic cell (for producing electrolysis), aeronautic gas cell (contained in a balloon), ecclesiastical cell (i.e., used to describe a

DOI: 10.1201/9781003332992-13

monastery or nunnery), or currently and definitely more commonly used to describe a cellphone.

You may have noticed that nowhere in this particular list are terms related to subject of this book: the science of electric vehicles. While some would argue that any kind of biological cell mechanism and/or part/organelle and cellphones are types of energy devices, producers, or consumers that can be renewed unless destroyed, it is not my intention to argue this point or issue either way; instead, it is my point that an important type of cell is not included in the above list—and that it should be included. We are referring to the electric cell, electrochemical cell, galvanic cell, voltaic cell—often referred to as a battery. This type of cell, no matter what we call it, is a device that generates electrical energy from chemical energy, usually consisting of two different conducting substances placed in an electrolyte. (Note that I could also include the solar cell in this discussion but I discussed these cells earlier in the texts.)

Before moving on to our basic discussion of fuel cells and their associated terminology and applications, it is important to point out that although fuel cells are not topics of discussion anywhere near as common as those other cells (cellphones), for instance, I predict that the day is coming (sooner than later) when we will refer to our fuel cells just as commonly as we mention our cellphones—for it will be the fuel cell that will power our lives, just as the cellphone powers and transmits our communication.

Key Terms

The key terms related to fuel cells are listed and defined in the glossary; however, to assist in understanding material presented in this chapter, the key components and operations related to fuel cells are defined in the following:

Alkaline fuel cell (AFC)—A type of hydrogen/oxygen fuel cell in which the electrolyte is concentrated potassium hydroxide (KOH) and the hydroxide ions (OH^-) are transported from the cathode to the anode.

Anion—A negatively charged ion; an ion that is attracted to the anode.

Anode—The electrode at which oxidation (a loss of electrons) takes place. For fuel cells and other galvanic cells, the anode is the negative terminal; for electrolytic cells (where electrolysis occurs), the anode is the positive terminal.

Bipolar plates—The conductive plate in a fuel cell stack that acts as a node for one cell and a cathode for the adjacent cell. The plate may be made of metal or a conductive polymer (which may be a carbon-filled composite). The plate usually incorporates flow channels for the fluid feeds and may also contain conduits for heat transfer.

Catalyst—A chemical substance that increases the rate of a reaction without being consumed; after the reaction, it can potentially be recovered from the reaction mixture and is chemically unchanged. The catalyst lowers the activation energy required, allowing the reaction to proceed more quickly or at a lower temperature. In a fuel cell, the catalyst facilitates the reaction of oxygen and hydrogen. It is usually made of platinum power very thinly coated onto carbon paper or cloth. The catalyst is rough and porous so the maximum surface area of the platinum can be exposed to the hydrogen or oxygen. The platinum-coated side of the catalyst faces the membrane in the fuel cell.

Catalyst Poisoning—The process of impurities binding to a fuel cell's catalyst (fuel cell poisoning), lowering the catalyst's ability to facilitate the desired chemical reaction.

Cathode—The electrode at which reduction (a gain of electrons) occurs. For fuel cells and other galvanic cells, the cathode is the positive terminal, for electrolytic cells (where electrolysis occurs) the cathode is the negative terminal.

Cation—A positively charged ion.

Combustion—The burning fire produced by the proper combination of fuel, heat, and oxygen. In the engine, the rapid burning of the air-fuel mixture that occurs in the combustion chamber.

Combustion chamber—In an internal combustion engine, the space between the top of the piston and the cylinder head in which the air-fuel mixture is burned.

Composite—Material created by combining materials differing in composition or form on a macroscale to obtain specific characteristics and properties. The constituents retain their identity; they can be physically identified, and they exhibit an interface among one another.

Compressed hydrogen gas (CHG)—Hydrogen gas compressed to a high pressure and stored at ambient temperature.

Compressed natural gas (CNG)—Mixtures of hydrocarbon gases and vapors, consisting principally of methane in gaseous form that has been compressed.

Compressor—A device used for increasing the pressure and density of gas.

Cryogenic liquefaction—The process through which gases such as nitrogen, hydrogen, helium, and natural gas are liquefied under pressure at very low temperatures.

Current collector—The conductive material in a fuel cell that collects electrons (on the anode side) or disburses electrons (on the cathode side). Current collectors are microporous (to allow fluid to flow through them) and lie in between the catalyst/electrolyte surfaces and the bipolar plates.

Direct methanol fuel cell (DMFC)—A type of fuel cell in which the fuel is methanol (CH_3OH) in gaseous or liquid form. The methanol is oxidized directly at the anode instead of first being reformed to produce hydrogen. The electrolyte is typically a PEM.

Dispersion—The spatial property of being scattered over an area or volume.

Electrode—A conductor through which electrons enter or leave an electrolyte. Batteries and fuel cells have a negative electrode (the anode) and a positive electrode (the cathode).

Electrolysis—A process that uses electricity, passing through an electrolytic solution of other appropriate medium, to cause a reaction that breaks chemical bonds (e.g., electrolysis of water to produce hydrogen and oxygen).

Electrolyte—A substance that conducts charged ions from one electrode to the other in a fuel cell, battery, or electrolyzer.

Endothermic—A chemical reaction that absorbs or requires energy (usually in the form of heat).

Energy content—Amount of energy for a given weight of fuel.

Energy density—Amount of potential energy in a given measurement of fuel.

Engine—A machine that converts heat energy into mechanical energy.

Ethanol (CH_3CH_2OH)—An alcohol containing two carbon atoms. Ethanol is a clear, colorless liquid and is the same alcohol found in beer, wine, and whiskey. Ethanol can be produced from cellulosic materials or by fermenting a sugar solution with yeast.

Exhaust emissions—Materials emitted into the atmosphere through any opening downstream of the exhaust ports of an engine, including water, particulates, and pollutants.

Exothermic—A chemical reaction that gives off heat.

Flammability limits—The flammability range of a gas is defined in terms of its lower flammability limit (LFL) and its upper flammability limit (UFL). Between the two limits is the flammable range in which the gas and air are in the right proportions to burn when ignited. Below the lower flammability limit, there is not enough fuel to burn. Above the high flammability limit, there is not enough air to support combustion.

Flashpoint—The lowest temperature under very specific conditions at which a substance will begin to burn.

Flexible fuel vehicle—A vehicle that can operate on a wide range of fuel blends (e.g., blends of gasoline and alcohol) that can be put in the same fuel tank.

Fuel—A material used to create heat or power through conversion in such processes as combustion or electrochemistry.

Fuel cell—A device that produces electricity through an electrochemical process, usually from hydrogen and oxygen.

Fuel cell poisoning—The lowering of a fuel cell's efficiency due to impurities in the fuel binding to the catalyst.

Fuel cell stack—Individual fuel cells connected in a series. Fuel cells are stacked to increase voltage.

Fuel processor—Device used to generate hydrogen from fuels such as natural gas, propane, gasoline, methanol, and ethanol for use in fuel cells.

Gas—Fuel gas such as natural gas, undiluted liquefied petroleum gases (vapor phase only), liquefied petroleum gas-air mixtures, or mixtures of these gases.

- **Natural gas**—Mixtures of hydrocarbon gases and vapors consisting principally of methane (CH_4) in gaseous form.
- **Liquefied petroleum gases (LPG)**—Any material composed predominately of any of the following hydrocarbons or mixtures of them: propane, propylene, butanes (normal butane or isobutane) and butylenes.
- **Liquefied petroleum gas-air mixture**—Liquefied petroleum gases distribute at relatively low pressures and normal atmospheric temperatures that have been diluted with air to produce desired heating value and utilization characteristics.

Gas diffusion—Mixing of two gases caused by random molecular motions. Gases diffuse very quickly, liquids diffuse much more slowly, and solids diffuse at a very slow (but often measurable) rate. Molecular collisions make diffusion slower in liquids and solids.

Graphite—Mineral consisting of a form of carbon that is soft, black, and lustrous and has a greasy feeling. Graphite is used in pencils, crucibles, lubricants, paints, and polishers.

Gravimetric energy density—Potential energy in a given weight of fuel.

Greenhouse effect—Warming of the earth's atmosphere due to gases in the atmosphere that allow solar radiation (visible, ultraviolet) to reach the earth's atmosphere but do not allow the emitted infrared radiation to pass back out of the earth's atmosphere.

Greenhouse gas (GHG)—Gases in the earth's atmosphere that contribute to the greenhouse effect.

Heat exchanger—Device (e.g., a radiator) that is designed to transfer heat from the hot coolant that flows through it to the air blown through it by the fan.

Heating value (Total)—The number of British thermal units (Btu) produced by the combustion of one cubic foot gas at constant pressure when the products of combustion are cooled to the initial temperature of the gas and air, when the water vapor formed during combustion is condensed, and when all the necessary corrections have been applied.

- **Lower (LHV)**—The value of the heat of combustion of a fuel measured by allowing all products of combustion to remain in the gaseous state. This method of measure does not consider the heat energy put into the vaporization of water (heat of vaporization).
- **Higher (HHV)**—The value of the heat of combustion of a fuel measured by reducing all of the products of combustion back to their original temperature and condensing all water vapor formed by combustion. This value considers the heat of vaporization of water.

Hybrid electric vehicle (HEV)—A vehicle combining a battery-powered electric motor with a traditional internal combustion engine. The vehicle can run on either the battery or the engine or both simultaneously, depending on the performance objectives for the vehicle.

Hydrides—Chemical compounds formed when hydrogen gas reacts with metals. Used for storing hydrogen gas.

Hydrocarbon (HC)—An organic compound containing carbon and hydrogen, usually derived from fossil fuels, such as petroleum, natural gas, and coal.

Hydrogen (H_2)—Hydrogen (H) is the most abundant element in the universe, but it is generally bonded to another element. Hydrogen gas (H_2) is a diatomic gas composed of two hydrogen atoms and is colorless and odorless. Hydrogen is flammable when mixed with oxygen over a wide range of concentrations.

Hydrogen-rich fuel—A fuel that contains a significant amount of hydrogen, such as gasoline, diesel fuel, methanol (CH_3OH), ethanol (CH_3CH_2OH), natural gas, and coal.

Impurities—Undesirable foreign material(s) in a pure substance or mixture.

Internal combustion engine (ICE)—An engine that converts the energy contained in a fuel inside the engine into motion by combusting the fuel. Combustion engines use the pressure created by the expansion of combustion product gases to do mechanical work.

Liquefied hydrogen (LH_2)—Hydrogen in liquid form. Hydrogen can exist in a liquid state but only at extremely cold temperatures. Liquid hydrogen typically has to be stored at −253°C (−423°F). The temperature requirements for liquid hydrogen storage necessitate expending energy to compress and chill the hydrogen into its liquid state.

Liquefied natural gas (LNG)—Natural gas in liquid form. Natural gas in a liquid at −162°C (−259°F) at ambient pressure.

Liquefied petroleum gas (LPG)—Any material that consists predominately of any of the following hydrocarbons or mixtures of hydrocarbons: propane, propylene, normal butane, isobutylene, and butylenes. LPG is usually stored under pressure to maintain the mixture in the liquid state.

Liquid—A substance that, unlike a solid, flows readily but, unlike a gas, does not tend to expand indefinitely.

Membrane—The separating layer in a fuel cell that acts as electrolyte (an ion exchanger) as well as a barrier film separating the gases in the anode and cathode compartments of the fuel cell.

Methanol (CH_3OH)—An alcohol containing one carbon atom. It has been used, together with some of the higher alcohols, as a high-octane gasoline component and is a useful automotive fuel.

Miles per gallon equivalent (MPGE)—Energy content equivalent to that of a gallon (114,320 Btu).

Molten carbonate fuel cell (MCFC)—A type of fuel cell that contains a molten carbonate electrolyte. Carbonate ions (CC_3^{-2}) are transported from the cathode to the anode. Operating temperatures are typically near 650°C.

Naflon®—Sulfonic acid in a solid polymer form that is usually the electrolyte of PEM fuel cells.

Natural gas—A naturally occurring gaseous mixture of simple hydrocarbon components (primarily methane) used as a fuel.

Nitrogen (N_2)—A diatomic colorless, tasteless, odorless gas that constitutes 78% of the atmosphere by volume.

Nitrogen oxides (NO_x)—Any chemical compound of nitrogen and oxygen. Nitrogen oxides result from high temperature and pressure in the combustion chambers of automotive engines and other power plants during the combustion process. When combined with hydrocarbons in the presence of sunlight, nitrogen oxides form smog. Nitrogen oxides are basic air pollutants; automotive exhaust emission levels of nitrogen oxides are regulated by law.

Oxidant—A chemical, such as oxygen, that consumes electrons in an electrochemical reaction.

Oxidation—Loss of one or more electrons by an atom, molecule, or ion.

Oxygen (O_2)—A diatomic colorless, tasteless, odorless gas that makes up about 21% of air.

Partial oxidation—Fuel reforming reaction where the fuel is oxidized partially to carbon monoxide and hydrogen rather than fully oxidized to carbon dioxide and water. This is accomplished by injecting air with the fuel stream prior to the reformer.

The advantage of partial oxidation over steam reforming of the fuel is that it is an exothermic reaction rather than an endothermic reaction and therefore generates its own heat.

Pascal (Pa)—The Pascal is the International System of Units (SI)-derived unit of pressure or stress. It is a measure of perpendicular force per unit area. It is equivalent to one newton per square meter.

Permeability—Ability of a membrane or other material to permit a substance to pass through it.

Phosphoric acid fuel cell (PAFC)—A type of fuel cell in which the electrolyte consists of concentrated phosphoric acid (H_3PO_4).

Polymer—Natural or synthetic compound composed of repeated links of simple molecules.

Polymer electrolyte membrane (PEM)—A fuel cell incorporating a solid polymer membrane used as its electrolyte. Protons (H^+) are transported from the anode to the cathode. The operating temperature range is generally 60°C–100°C.

Polymer electrolyte membrane fuel cell (PEMFC or PEFC)—A type of acid-based fuel cell in which the transport of protons (H^+) from the anode to the cathode is through a solid, aqueous membrane impregnated with an appropriate acid. The electrolyte is a called a polymer electrolyte membrane (PEM). The fuel cells typically run at low temperatures (<100°C).

Reactant—A chemical substance that is present at the start of a chemical reaction.

Reactor—Device or process vessel in which chemical reactions (e.g., catalysis in fuel cells) take place.

Reformate—Hydrocarbon fuel that has been processed into hydrogen and other products for use in fuel cells.

Reformer—Device used to generate hydrogen from fuels such as natural gas, propane, gasoline, methanol, and ethanol for use in fuel cells.

Reforming—A chemical process in which hydrogen-containing fuels react with steam, oxygen, or both to produce a hydrogen-rich gas stream.

Reformulated gasoline—Fuel (gasoline) that is blended so that, on average, it reduces volatile organic compounds and toxic air emissions significantly relative to conventional gasolines.

Regenerative fuel cell—A fuel cell that produces electricity from hydrogen and oxygen and can use electricity from solar power or some other source to divide the excess water into oxygen and hydrogen fuel to be reused by the fuel cell.

Solid oxide fuel cell (SOFC)—A type of fuel cell in which the electrolyte is a solid, nonporous metal oxide, typically zirconium oxide (ZrO_2) treated with Y_2O_3 and O^{-2}, that is transported from the cathode to the anode. Any CO in the reformate gas is oxidized to CO_2 at the anode. Temperatures of operation are typically 800°C–1,000°C.

Sorbent—Material that sorbs another (i.e., has the capacity or tendency to take it up either by adsorption or absorption).

Sorption—Process by which one substance takes up or holds another.

Steam reforming—The process for reacting a hydrocarbon fuel, such as natural gas, with steam to produce hydrogen as a product. This is a common method for bulk hydrogen generation.

Turbine—Machine for generating rotary mechanical power from the energy in a stream of fluid. The energy, originally in the form of heat or pressure energy, is converted to velocity energy by passing through a system of stationary and moving blades in the turbine.

Turbocharger—A device used for increasing the pressure and density of a fluid entering a fuel cell power plant using a compressor driven by a turbine that extracts energy form the exhaust gas.

Turbocompressor—Machine for compressing air or other fluids (react if supplied to a fuel cell system) in order to increase the reactant pressure and concentration.

Volumetric energy density—Potential energy in a given volume of fuel.

Watt (W)—A unit of power equal to one Joule of work performed per second: 746 watts is the equivalent of one horsepower. The watt is named for James Watt, Scottish engineer (1736–1819) and pioneer in steam engine design.

Wt.%—The term wt.% (abbreviation for weight percent) is widely used in hydrogen storage research to denote the amount of hydrogen stored on a weight basis, and the term mass % is also occasionally used. The term can be used for materials that store hydrogen or for the entire storage system (e.g., material or compressed/liquid hydrogen as well as the tank and other equipment required to contain the hydrogen such as insulation, valves, regulators, etc.). For example, 6 wt.% on a system basis means that 6% of the entire system by weight is hydrogen. On a material basis, the wt.% is the mass of hydrogen divided by the mass of material plus hydrogen.

HYDROGEN FUEL CELLS AND OTHER TYPES[1]

Before presenting a detailed discussion of hydrogen fuel cells and other fuel cells it is important to list a few general facts about the cells. Note that until recently not that many people had heard about hydrogen and fuel cells, but these technologies are bursting on to the scene and have the potential to solve some of the biggest problems in energy ranging from commercial buildings to transportation. The truth be told most people are more familiar with solar, wind, and battery power but the fuel cell technologies could add to the country's diverse energy mix. To set the stage for the material to follow here are a few things to know about hydrogen and fuel cells.

- Hydrogen is the most abundant element on earth. Moreover, hydrogen is an alternative fuel that has very high energy content by weight. And note that enormous quantities of hydrogen are locked up in water, hydrocarbons, and other organic matter. Hydrogen can be produced from assorted, domestic resources including fossil fuels, biomass (i.e., plant-based material), and water electrolysis with wind, solar, or grid electricity. Note that the environmental impact and energy efficiency of hydrogen depends on how it is produced.
- Hydrogen and fuel cells can be used to power in a broad range of applications, including powering buildings, cars, trucks, to portable electronic

[1] Information in this section is from USDOE (2008) Hydrogen, Fuel Cells & Infrastructure Technologies Program. Accessed @ http://www1.eere.energy.gov/hydrogenandfuelcells/production/basics.html.

devices and backup power systems. Fuel cells are an attractive option for critical load functions such as data centers, telecommunications towers, hospitals, emergency response systems, and even military applications for national defense because they are grid independent.

- Fuel cells do not produce emissions from combustion of hydrocarbon fuels, making them attractive like battery-powered vehicles-on-wheels. Unlike batteries, fuel cells do not run down or need to recharge; that is, instead, as long as there is a constant source of fuel and oxygen. Compared to conventional gasoline and diesel vehicles, fuel cell vehicles can even reduce carbon dioxide by up to half if the hydrogen is produced by natural gas and by 90% if the hydrogen is produced by renewable energy, such as wind and solar. Only water is emitted from the tailpipe and no pollutants.
- Fuel cell-powered vehicles-on-wheels are similar to today's gasoline cars, at the present time fuel cell electric vehicles can have a driving range of more than 300 miles on one tank of hydrogen fuel. The good news is that they can refuel in just a few minutes and the fueling experience is almost identical to a gas station. Since the fuel cell engine has no moving parts, you'll never need to change the oil. Moreover, because a fuel cell is more than twice as efficient as an internal combustion engine, a fuel cell car travels farther on that tank of hydrogen than a traditional vehicle would on gasoline. This means you only need approximately half the amount of hydrogen with double the fuel economy.
- Hydrogen fueling stations are increasing throughout the country, with California having more than 30 retail hydrogen fueling stations in operation, and plans are in place to increase the number to more than 100 stations. In the northeast area of the United States there are several stations ready to open up. The current plan to install hydrogen in existing gas stations is the result of careful planning (and common sense, which is similar to practical thinking). These efforts are giving hydrogen customers the confidence that they will be able to find a station and access the hydrogen in the same manner they would purchase gasoline or diesel fuels. Other markets are expected to develop as customer demand increases.

Containing only one electron and one proton, hydrogen, chemical symbol H, is the simplest element on earth. Hydrogen is a diatomic molecule—each molecule has two atoms of hydrogen (which is why pure hydrogen is commonly expressed as H_2). Although abundant on earth as an element, hydrogen combines readily with other elements and is almost always found as part of another substance, such as water hydrocarbons, or alcohols. Hydrogen is also found in biomass, which includes all plants and animals.

- Hydrogen is an energy carrier, not an energy source. Hydrogen can store and deliver usable energy, but it doesn't typically exist by itself in nature; it must be produced from compounds that contain it.
- Hydrogen can be produced using diverse, domestic resources including nuclear; natural gas and coal; and biomass and other renewables including

solar, wind, hydroelectric, or geothermal energy. This diversity of domestic energy sources makes hydrogen a promising energy carrier and important to our nation's energy security. It is expected and desirable for hydrogen to be produced using a variety or resources and process technologies (or pathways).

- DOE focuses on hydrogen production technologies that result in near-zero, net greenhouse gas emissions and use renewable energy sources, nuclear energy, and coal (when combined with carbon sequestration). To ensure sufficient clean energy for our overall energy needs, energy efficiency is also important.
- Hydrogen can be produced via various process technologies, including thermal (natural gas reforming, renewable liquid and bio-oil processing, and biomass and coal gasification), electrolytic (water splitting using a variety of energy resources), and photolytic (splitting water using sunlight via biological and electrochemical materials).
- Hydrogen can be produced in large, central facilities (50–300 miles from point of use), smaller semi-central (located within 25–100 miles of use), and distributed (near or at point of use). Learn more about distributed vs. centralized production.
- In order for hydrogen to be successful in the marketplace, it must be cost-competitive with the available alternatives. In the light-duty vehicle transportation market, this competitive requirement means that hydrogen needs to be available untaxed at $2-$3/gge (gasoline gallon equivalent). This price would result in hydrogen fuel cell vehicles having the same cost to the consumer on a cost-per-mile-driven basis as a comparable conventional internal combustion engine or hybrid vehicle.
- DOE is engaged in research and development of a variety of hydrogen production technologies. Some are further along in development than others—some can be cost-competitive for the transition period (beginning in 2015), and others are considered long-term technologies (cost-competitive after 2030).

Infrastructure is required to move hydrogen from the location where it's produced to the dispenser at a refueling station or stationary power site. Infrastructure includes the pipelines, trucks, railcars, ships, and barges that deliver fuel, as well as the facilities and equipment needed to load and unload them.

Delivery technology for hydrogen infrastructure is currently available commercially, and several US companies deliver bulk hydrogen today. Some of the infrastructure is already in place because hydrogen has long been used in industrial applications, but it's not sufficient to support widespread consumer use of hydrogen as an energy carrier. Because hydrogen has a relatively low volumetric energy density, its transportation, storage, and final delivery to the point of use comprise a significant cost and result in some of the energy inefficiencies associated with using it as an energy carrier.

Options and trade-offs for hydrogen delivery from central, semi-central, and distributed production facilities to the point of use are complex. The choice of a hydrogen production strategy greatly affects the cost and method of delivery; for example, larger, centralized facilities can produce hydrogen at relatively low costs due to economies of scale, but the delivery costs for centrally produced hydrogen are higher than the delivery costs for semi-central or distributed production options (because the point of use is farther away). In comparison, distributed production facilities have relatively low delivery costs, but the hydrogen production costs are likely to be higher—lower volume production means higher equipment costs on a per-unit-of-hydrogen basis.

Key challenges to hydrogen delivery include reducing delivery cost, increasing energy efficiency, maintaining hydrogen purity, and minimizing hydrogen leakage. Further research is needed to analyze the trade-offs between the hydrogen production options and the hydrogen delivery options taken together as a system. Building a national hydrogen delivery infrastructure is a big challenge. It will take time to develop and will likely include combinations of various technologies. Delivery infrastructure needs and resources will vary by region and type of market (e.g., urban, Interstate, or rural). Infrastructure options will also evolve as the demand for hydrogen grows and as delivery technologies develop and improve.

Hydrogen Storage

Storing enough hydrogen on-board a vehicle to achieve a driving range of greater than 300 miles is a significant challenge. On a weight basis, hydrogen has nearly three times the energy content of gasoline (120 MJ/kg for hydrogen vs. 44 MJ/kg for gasoline). However, on a volume basis the situation is reversed (8 MJ/L for liquid hydrogen vs. 32 MJ/L for gasoline). On-board hydrogen storage in the range of 5–13 kg H_2 is required to encompass the full platform of light-duty vehicles. Hydrogen can be stored in a variety of ways, but for hydrogen to be a competitive fuel for vehicles, the hydrogen vehicle must be able to travel a comparable distance to conventional hydrocarbon-fueled vehicles. Hydrogen can be physically stored as either a gas or a liquid. Storage as a gas typically requires high-pressure tanks (5,000–10,000 psi tank pressure). Storage of hydrogen as a liquid requires cryogenic temperatures because the boiling point of hydrogen at one atmosphere pressure is −252.8°C. Hydrogen can also be stored on the surfaces of solids (by adsorption) or within solids (by absorption). In adsorption, hydrogen is attached to the surface of material either as hydrogen molecules or as hydrogen atoms. In absorption, hydrogen is dissociated into H-atoms, and then the hydrogen atoms are incorporated into the solid lattice framework. Hydrogen storage in solids may make it possible to store large quantities of hydrogen in smaller volumes at low pressures and at temperatures close to room temperature. It is also possible to achieve volumetric storage densities greater than liquid hydrogen because the hydrogen molecule is dissociated into atomic hydrogen within the metal hydride lattice structure. Finally, hydrogen can be stored through the reaction of hydrogen-containing materials with water (or other compound such as alcohols). In this case, the hydrogen is effectively stored in both the material and in water. The term "chemical hydrogen storage" or chemical hydrides is used to describe this form of hydrogen storage. It is also possible to store hydrogen in the chemical structures of liquids and solids.

DID YOU KNOW?

Hydrogen fuel cell vehicles (FCVs) emit approximately the same amount of water per mile as vehicles using gasoline-powered internal combustion engines (ICEs).

How a Hydrogen Fuel Cell Works

A basic fuel cell (see Figure 13.1) is an electrochemical device that uses the chemical energy of hydrogen to cleanly and efficiently produce electricity with water and heat as by-products. Fuel cells are unique in terms of variety of their potential applications; they can provide energy for systems as large as a utility power station and as small as a laptop computer. Fuel cells have several benefits over conventional combustion-based technologies currently used in many power plants and passenger vehicles. They produce much smaller quantities of greenhouse gases and none of the air pollutants that create smog and cause health problems. If pure hydrogen is used as a fuel, fuel cells emit only heat and water as by-products.

A hydrogen fuel cell is a device that uses hydrogen (or hydrogen-rich fuel) and oxygen to create electricity by an electrochemical process. A single fuel cell consists of an electrolyte and two catalyst-coated electrodes (a porous anode and cathode). While there are different fuel cell types, all fuel cells work similarly:

- Hydrogen, or a hydrogen-rich fuel, is fed to the anode where a catalyst separates hydrogen's negatively charged electrons from positively charge ions (protons).
- At the cathode, oxygen combines with electrons and, in some cases, with species such as protons or water, resulting in water or hydroxide ions, respectively.
- For polymer electrolyte membrane and phosphoric acid fuel cells, protons move through the electrolyte to the cathode to combine with oxygen and electrons, producing water and heat.
- For alkaline, molten carbonate, and solid oxide fuel cells, negative ions travel through the electrolyte to the anode where they combine with hydrogen to generate water and electrons.

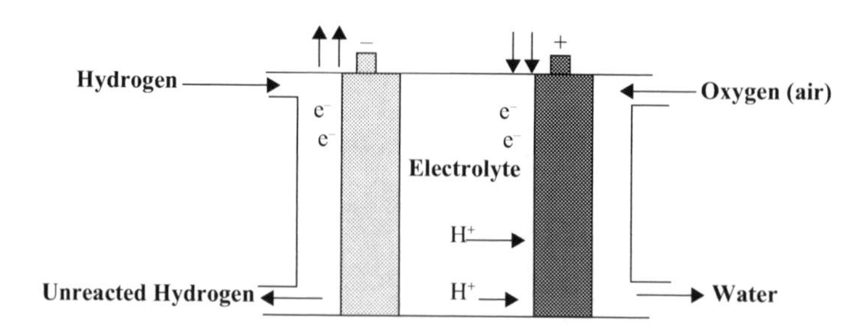

FIGURE 13.1 Fuel cell structure.

- The electrons from the anode cannot pass through the electrolyte to the positively charged cathode; they must travel around it via an electrical circuit to reach the other side of the cell. This movement of electrons is an electrical current.

Hydrogen has unique physical and chemical properties, which present benefits and challenges to its successful widespread adoption as a fuel. Hydrogen is the lightest and smallest element in the universe. Hydrogen is 14 times lighter than air and rises at a speed of almost 20 m/s, six times faster than natural gas, which means that when released, it rises and disperses quickly. Hydrogen is also odorless, colorless, and tasteless, making it undetectable by human senses. For these reasons, hydrogen systems are designed with ventilation and leak detection. Natural gas is also odorless, colorless, and tasteless, but a sulfur-containing odorant is added so people can detect it. There is no known odorant light enough to "travel with" hydrogen at an equal dispersion rate, so odorants are not used to provide a detection method. Many odorants can also contaminate fuel cells.

OTHER TYPES OF FUEL CELLS[2]

In addition to hydrogen fuel cells, there are other types of fuel cells. These other types are classified primarily by the kind of electrolyte they employ. The classification determines the kind of electrochemical reactions that take place in the cell, the kind of catalysts required, the temperature range in which the cell operates, the fuel required, and other factors. These characteristics, in turn, affect the applications for which these cells are most suitable. There are several types of fuel cells currently under development; those with the most promise, at the present time, are discussed below.

Direct Methanol Fuel Cells

Most fuel cells are powered by hydrogen, which can be fed to the fuel cell system directly or can be generated within the fuel cell system by reforming hydrogen-rich fuels such as methanol, ethanol, and hydrocarbon fuels. Direct methanol fuel cells (DMFCs), however, are powered by pure methanol, which is usually mixed with water and fed directly to the fuel cell anode.

Direct methanol fuel cells do not have many of the fuel storage problems typical of some fuel cell systems because methanol has a higher energy density than hydrogen—though less than gasoline or diesel fuel. Methanol is also easier to transport and supply to the public using our current infrastructure because it is a liquid, like gasoline. DMFCs are often used to provide power to portable fuel cell applications such as cellphones or laptop computers.

[2] From USDOE (2015) *Types of Fuel Cells*. United States Department of Energy. Accessed 7/7/22 @ http://energy.gov/eere/fuelcells/types-fuel-cells

Alkaline Fuel Cells

Alkaline fuel cells (AFCs) were one of the first fuel cell technologies developed, and they were the first type widely used in the US space program to produce electrical energy and water on-board spacecraft. These fuel cells use a solution of potassium hydroxide in water as the electrolyte and can use a variety of non-precious metals as a catalyst at the anode and cathode. High-temperature AFCs operate at temperatures between 100°C and 250°C (212°F and 482°F). However, newer AFC designs operated at lower temperatures of roughly 23°C–70°C (74°F–158°F). In recent years, novel AFCs that use a polymer membrane as the electrolyte have been developed. These fuel cells are closely related to conventional PEM fuel cells, except that they use an alkaline membrane instead of an acid membrane. The high performance of AFCs is due to the rate at which electrochemical reactions take place in the cell. They have also demonstrated efficiencies above 60% in space applications.

The disadvantage of this fuel cell type is that it is easily poisoned by carbon dioxide (CO_2). In fact, even the small amount of CO_2 in the air can affect this cell's operation, making it necessary to purify both the hydrogen and oxygen used in the cell. This purification process is costly. Susceptibility to poisoning also affects the cell's lifetime (the amount of time before it must be replaced), further adding to cost. Alkaline membrane cells have lower susceptibility to CO_2 poisoning than liquid electrolyte AFCs do, but performance still suffers as a result of CO_2 that dissolves into the membrane.

Cost is less of a factor for remote locations, such as in space or under the sea. However, to compete effectively in most mainstream commercial markets, these fuel cells will have to become more cost-effective. To be economically viable in large-scale utility applications, AFCs need to reach operating times exceeding 40,000 hours, something that has not yet been achieved due to material durability issues. This obstacle is possibly the most significant in commercializing this fuel cell technology.

Phosphoric Acid Fuel Cells

Phosphoric acid fuel cells (PAFCs) use liquid phosphoric acid as an electrolyte—the acid is contained in a Teflon-bonded silicon carbide matrix—and porous carbon electrodes containing a platinum catalyst.

The PAFC is considered the "first generation" of modern fuel cells. It is one of the most mature cell types and the first to be used commercially. This type of fuel cell is typically used for stationary power generation, but some PAFCs have been used to power large vehicles such as city buses.

PAFCs are more tolerant of impurities in fossil fuels that have been reformed into hydrogen than PEM cells, which are easily "poisoned" by carbon monoxide because carbon monoxide binds to the platinum catalyst at the anode, decreasing the fuel cell's efficiency. PAFCs are more than 85% efficient when used for the co-generation of electricity and heat but they are less efficient at generating electricity alone (37%–42%). PAFC efficiency is only slightly more than that of combustion-based power plants, which typically operate at around 33% efficiency. PAFCs are also less powerful than other fuel cells, given the same weight and volume. As a result, these fuel cells are typically large and heavy. PAFCs are also expensive. They require much higher loadings of expensive platinum catalyst than other types of fuel cells do, which raises the cost.

Molten Carbonate Fuel Cells

Molten carbonate fuel cells (MCFCs) are currently being developed for natural gas and coal-based power plants for electrical utility, industrial, and military applications. MCFCs are high-temperature fuel cells that use an electrolyte composed of a molten carbonate salt mixture suspended in a porous, chemically inert ceramic lithium aluminum oxide matrix. Because they operate at high temperatures of 650°C (roughly 1,200°F), non-precious metals can be used as catalysts at the anode and cathode, reducing costs.

Improved efficiency is another reason MCFCs offer significant cost reductions over phosphoric acid fuel cells. Molten carbonate fuel cells, when coupled with a turbine, can reach efficiencies approaching 65%, considerably higher than the 37%–42% efficiencies of a phosphoric acid fuel cell plant. When the water heat is captured and used, overall fuel efficiencies can be over 85%.

Unlike alkaline, phosphoric acid, and PEM fuel cells, MCFCs do not require an external reformer to convert fuels such as natural gas and biogas to hydrogen. At the high temperatures at which MCFCs operate, methane and other light hydrocarbons in these fuels are converted to hydrogen within the fuel cell itself by a process called internal reforming.

The primary disadvantage of current MCFC technology is durability. The high temperatures at which these cells operated and the corrosive electrolyte used accelerate component breakdown and corrosion, decreasing cell life. Scientists are currently exploring corrosion-resistant materials for components as well as fuel cell designs that double cell life from the current 40,000 hours (~5 years) without decreasing performance.

Solid Oxide Fuel Cells

Solid oxide fuel cells (SOFCs) use a hard, nonporous ceramic compound as the electrolyte. SOFCs are around 60% efficient at converting fuel to electricity in applications designed to capture and utilize the system's waste heat (co-generation), overall fuel use efficiencies could top 85%.

SOFCs operate at very high temperatures—as high as 1,000°C (1,830°F). High-temperature operation removes the need for precious metal catalyst, thereby reducing cost. It also allows SOFCs to reform fuels internally, which enables the use of a variety of fuels and reduces the cost associated with adding a reformer to the system.

SOFCs are also the most sulfur-resistant fuel cell type; they can tolerate several orders of magnitude more sulfur than other cell types can. In addition, they are not poisoned by carbon monoxide, which can even be used as fuel. This property allows SOFCs to use natural gas, biogas, and gases made from coal. High-temperature operation has disadvantages. It results in a slow start-up and requires significant thermal shielding to retain heat and protect personnel, which may be acceptable for utility applications but not for transportation. The high operating temperatures also place stringent durability requirements on materials. The development of low-cost materials with high durability at cell operating temperatures is the key technical challenge facing this technology.

Scientists are currently exploring the potential for developing lower-temperature SOFCs operating at or below 700°C that have fewer durability problems and cost less. Lower-temperature SOFCs have not yet matched the performance of the higher temperature systems, however, and stack materials that will function in this lower temperature range are still under development.

Reversible Fuel Cells

Reversible fuel cells produce electricity from hydrogen and oxygen and generate heat and water as by-products, just like other fuel cells. However, reversible fuel cell systems can also use electricity from solar power, wind power, or other sources to split water into oxygen and hydrogen fuel through a process called electrolysis. Reversible fuel cells can provide power when needed, but during times of high-power production form other technologies (such as when high winds lead to an excess of available wind power), reversible fuel cells can store the excess energy in the form of hydrogen. This energy storage capability could be a key enabler for intermittent renewable energy technologies.

Polymer Electrolyte Membrane Fuel Cells (PEM)

Polymer electrolyte membrane (PEM)—also called proton exchange membrane—fuel cells deliver high power density and offer the advantages of low weight and volume compared with other fuel cells. PEM fuel cells use a solid polymer as an electrolyte and porous carbon electrodes contain platinum or platinum alloy catalyst. They need only hydrogen, oxygen from the air, and water to operate. They are typically fueled with pure hydrogen supplied from storage tanks or reformers. PEM fuel cells operate at relatively low temperatures, around 80°C (176°F). Low-temperature operation allows them to start quickly (less warm-up time, which makes PEM fuel cells particularly suitable for use in passenger vehicles) and results in less wear on system components, resulting in better durability. However, it requires that a noble-metal catalyst (typically platinum) be used to separate the hydrogen's electrons and protons, adding to system cost. The platinum catalyst is also extremely sensitive to carbon monoxide poisoning, making it necessary to employ an additional reactor to reduce carbon monoxide in the fuel gas if the hydrogen is derived from a hydrocarbon fuel. This reactor also adds cost. PEM fuel cells are used primarily for some stationary applications and transportation applications.

Parts of a PEM Fuel Cell[3]

The PEM fuel cell just mentioned is the current focus of research for fuel cell vehicle applications—for vehicles-on-wheels. PEM cells are made from several layers of different materials.

The main parts of a **Polymer Electrolyte Membrane Fuel Cells (PEM)** are:

Polymer electrolyte membrane (PEM)—also called proton exchange membrane—fuel cells deliver high power density and offer the advantages of low weight and volume compared with other fuel cells. PEM fuel cells use a solid polymer as an electrolyte and porous carbon electrodes contain platinum or platinum alloy catalyst. They need only hydrogen, oxygen from the air, and water to operate. They are typically fueled with pure hydrogen supplied from storage tanks or reformers. PEM fuel cells operate at relatively low temperatures, around 80°C (176°F). Low-temperature operation allows them to start quickly (less warm-up time, which makes PEM fuel cells particularly suitable for use in passenger vehicles) and results in less wear on system components, resulting in better durability. However, it requires that a noble-metal catalyst (typically platinum) be used to separate the hydrogen's electrons and protons, adding to system cost. The platinum catalyst is also extremely sensitive to carbon monoxide poisoning, making it necessary to employ an additional reactor to reduce carbon monoxide in the fuel gas if the hydrogen is derived from a hydrocarbon fuel. This reactor also adds cost. PEM fuel cells are used primarily for some stationary applications and transportation applications. PEM fuel cells are made from several layers of different materials. The heart of a PEM fuel cell is the membrane electrode assembly (MEA), which includes the membrane, the catalyst layers, and gas diffusion layers (GDLs). These parts and others of the PEM fuel cell are described below. These other parts include hardware components including gaskets, which provide a seal around the MEA to prevent league of gases, and bipolar plates, which are used to assemble individual PEM fuel cells into a fuel cell stack and provide channels for the gaseous fuel and air.

Membrane Electrode Assembly

The membrane electrode assembly (MEA) of a PEM fuel cell is formed by the membrane, catalyst layers (anode and cathode), and diffusion media.

Polymer Electrolyte Membrane

Looking like ordinary kitchen plastic wrap the polymer electrolyte membrane, or PEM (aka proton exchange membrane), is a specially treated material that conducts only positively charged ions and blocks the electrons. Because it must permit only the necessary ions to pass between the anode and cathode, the PEM is the key to the fuel cell technology. The chemical reaction would be disrupted if other substances are allowed to pass through the electrolyte. The membrane is manufactured thin—in some cases under 20 microns—for use in transportation applications.

---

[3] This section is based on EERE (2022) *Parts of a Fuel Cell* Accessed 7/7/22 @ https://www.energy.gov/eere/feulecells/parts-fuel-cell.

Catalyst Layers

To both sides of the membrane a layer of catalyst is added (i.e., to the anode layer side and the cathode layer side). Conventional catalyst layers include nanometer-sized particles of platinum dispersed on a high-surface-area carbon support. The supported platinum catalyst is mixed with an ion-conducting ionomer polymer (composed of repeat units of both electrically neutral repeat units and ionized units covalently bonded to the polymer backbone as pendant group of shared moieties) and sandwiched between the membrane and the GDLs (gas diffusion layers). Note that on the anode side the platinum catalyst enables hydrogen molecules to be split into protons and electrons. Now on the other side, the cathode side, the platinum catalyst enables oxygen reduction by reacting with the protons generated by the anode, producing water. It is the ionomer mixed into the catalyst layers that allows the protons to travel through these layers.

Gas Diffusion Layers (GDLs)

Outside the catalyst layers sit the GDLs which facilitate transport of reactants into the catalyst layer, as well as removal of the product water. Typically composed of a sheet of carbon paper in which the carbon fibers and partially coated with polytetrafluoroethylene (PTFE) is what the GDL all about, structurally. This structure allows the gases to diffuse rapidly through the pores in the CDL. Note these pores are kept open by hydrophobic PTFE (i.e., repels water), which prevents excessive water buildup in many cases, the inner surface of the GDL is coated with a thin layer of higher-surface-area carbon mixed with PRFE (a microporous layer); it's all about water retention (needed to maintain membrane conductivity) and water release (needed to keep the pore open so hydrogen and oxygen can diffuse into the electrodes).

Hardware Components

MEAs are the heart of the fuel cell and where power is product, but it does not function alone; it requires hardware components required to enable effective MEA operation. The hardware components of concern to us are the bipolar plates and gaskets.

Bipolar Plates

Under normal operating conditions, individual MEAs produce less than 1 V. The problem is that most applications where MEAs are used require higher voltages. So, the key is to stack the MEAs. Stacking MEAs on top of each other connected in series provides a usable output voltage. It is a matter of sandwiching each cell into the stack between two bipolar plates to separate it from neighboring cells. These plates made of metal, carbon, or composites provide electrical conduction between cells, as well as providing physical strength to the stack. In the plates a flow field is inserted; this is a series of channels that are machined or stamped into the plate to allow gases for flow over the MEA. Additional channels inside each plate may be used to circulate a liquid coolant.

Stacking of bipolar plates is discussed in the next section.

Gaskets

Gaskets are sandwiched between two bipolar plates as each MEA in a fuel cell stack requires gaskets; they are needed around the edges of the MEA to make a gas-tight seal. These gaskets are usually made of a rubbery polymer.

FUEL CELL SYSTEMS

Design of fuel cell systems is complex and can vary significantly depending on the cell type and application. However, in this discussion the focus is on fuel cell stacks and fuel processors.

Fuel Cell Stack

As mentioned the fuel cell stack is the heart of a fuel cell power system. The electrochemical reactions within the fuel cell generates direct current (DC) electricity. Because a single cell produces less than 1 V, it is insufficient for most applications. So, individual fuel cells are typically combined in series into a fuel cell stack. Actually, a typical fuel cell stack can consist of hundreds of fuel cells. So, how much power can be generated by the fuel cell stack? It depends, of course, on several factors, such as fuel cell type, cell size, the temperature at which it operates, and the pressure of the gases supplied to the cell.

Fuel Processor

Another important component found in many fuel cell systems is the fuel processor. This component converts fuel into a form usable by the fuel cell. The type of fuel processor used is dependent on the type of fuel and type of fuel cell, it can be as simple as a sorbent bed to remove impurities, of a combination of multiple reactors and sorbents.

If a system is powered by a conventional fuel that is hydrogen rich, such as methanol, gasoline, diesel, or gasified coal, a device called a *reformer* is often used to convert hydrocarbons into a gas mixture of hydrogen and carbon compounds called "reformate" (a gasoline blending stock). In some systems, the reformate is then sent to a set of reactors to convert carbon monoxide to carbon dioxide and remove any trace amount of carbon monoxide remaining and a sorbent bed to remove other impurities, such as sulfur compounds, before it is sent to the fuel cell stack. This process prevents impurities in the gas from binding with the fuel cell catalysts. Note that this binding process is called "poisoning" because it reduces the efficiency and life expectancy of the fuel cell.

Molten carbonate and solid oxide fuel cells operate at temperatures high enough that fuel can be reformed in the fuel cell itself. This process is called internal reforming. Fuel cells that use internal reforming still need traps to remove impurities from the unreformed fuel before it reaches the fuel cell. Carbon dioxide is released from both internal and external reforming. However, due to the fuel cells' high efficiency, less carbon dioxide is emitted than by internal combustion engines, such as those used in gasoline-powered vehicles.

14 Economy and Range

INTRODUCTION

One of the main selling points for purchasing an electric or fuel cell vehicle-on-wheels is economy (perceived or actual)—saving fuel (hybrids) and money. One of the limiting factors about purchasing an electric or fuel cell vehicle is range—electric vehicle must be recharged and that is inconvenient and time consuming for long-distance trips, while gas-powered vehicles are convenient and relatively easy to fuel up and to get back on the road.

So, what is needed here right now is to provide the advantages and disadvantages of electric vehicles-on-wheels.

Let's begin with the advantages of electric vehicles-on-wheels. First, electric vehicles are environmentally friendly—they do not pollute the air as do conventional vehicles. Second, electric vehicles run on renewable energy—conventional vehicles work on the burning fossil fuels that exhaust fossil fuel reserves on earth. Third, they are cost-effective (at the present time) because electricity is much cheaper than fuels like petrol (gasoline) and diesel, which suffer a frequent price hike (sounds familiar?). If solar or wind power is used at home the recharging of batteries is cost-effective. Fourth, electric vehicles have fewer moving parts so wear and tear is less as compared to conventional auto parts. Moreover, at the present time repair work is also simple and less expensive relative to combustion engines. Fifth, electric vehicles give a much smoother driving experience. The absence of rapidly moving parts makes them much quieter with low sound generation. Finally, governments in various countries have offered tax credits as an incentive to encourage people to purchase and use electric vehicles as a go-green initiative.

The disadvantages of electric vehicles begin first with their high initial cost. Electric vehicles are still very expensive and many consumers consider them not as affordable as conventional vehicles. Second, have you ever tried to find a vehicle charging station? Good luck if you are looking. The truth be known: people need to drive long distances and they worry about suitable charging stations during their trip and more often than not find that few if any charging stations are presently available. Third, unlike conventional vehicles that require a few minutes for refilling fuel, recharging of the electric vehicle takes much more time, which is generally a few hours. Fourth, presently there aren't too many electric models to choose from when it comes to the looks, designs, or customized versions. Lastly, electric vehicles have less driving range—this is one of the reasons purchasers are reluctant to purchase electric vehicles. At the present time the driving range of the electric vehicle is found to be less as compared to conventional vehicles. Electric vehicles can be suitable for day-to-day travel but can be problematic for a long-distance journey.

After reviewing the advantages and disadvantages the obvious question might be: "How do we get over or modify or delete electrical vehicles?"

DOI: 10.1201/9781003332992-14

Good question and the solution is complex but not impossible to solve—every problem or shortcoming has a solution. The solution to range problems with electric vehicles is also complex but later in this chapter two "modest proposals," of sorts, offer potential solutions to the range problem. However, for now it is necessary to discuss in detail the fuel economy and range of all-electric vehicles, fuel cell vehicles, plug-in Hybrid Electric Vehicles, and ethanol flexible fuel vehicles (i.e., vehicles that can run on gasoline or 85% ethanol-gasoline blend). The best source of this information is both the US Environmental Protection Agency (USEPA) and the US Department of Energy (DOE) Fuel Economy Guide (2022) available on the web at *fueleconomy.gov*. Note: Table 14.1 is adapted from this source.

ALL-ELECTRIC VEHICLES

As mentioned, all-electric vehicles (EVs) are propelled by one of more electric motors powered by a rechargeable battery. Note that EVs are energy-efficient and emit no tailpipe pollutants; however, keep in mind that the power plant producing the electricity may emit pollution.

Electric motors have several performance advantages and benefits. They are quiet, have instant torque for quick acceleration, enable regenerative braking, and require less maintenance than internal combustion engines.

Current EVs (i.e., 2022) typically have a shorter driving range than comparable gasoline or hybrid vehicles, and their range is more sensitive to driving style, driving conditions, and accessory use. Fully recharging the battery can take several hours—though a "fast charge" to 80% capacity may take as little as 30 minutes. Currently, charging options outside the home are expanding and as of 2022 there are more than 43,000 public and 1,000 workplace stations available. At the present time EVs are more expensive than comparable conventional vehicles and hybrids due to the cost of the large battery. Note that as researchers develop vehicle power innovations and manufacturers continue to improve the driving range and reduce the cost of these vehicles, they are becoming more practical and affordable for a wider range of consumers.

In Table 14.1 selected EVs are listed with parameters showing their fuel economy (combination of city and highway operation), driving range, and charging time in hours @ 240 V.

FUEL CELL VEHICLES

Fuel cell vehicles (FCVs) are not quite ready for the mass market, nor are there enough refueling stations nationwide in the United States. However, a few fuel cell vehicles are available for lease and sale in select markets of the country, mostly in California. FEVs are propelled by electrical motors powered by fuel cells, which produce electricity from the chemical energy of hydrogen. With regard to efficiency, fuel cell technology is more efficient than internal combustion engines and environmentally cleaner—the only by-product of a hydrogen fuel cell vehicle is water.

TABLE 14.1
Selected EV Models and Fuel Economy/Driving/Charging Parameters

Model	Motor	Fuel Economy (MPGe)	Driving Range (comb city/hwy)	Charge Time (hrs @ 240V)
Two-Seater Cars	820kW AC	NA	NA	NA
		Subcompact Cars		
BMW	250kW EESM	109	301	10
		Compact Cars		
Porsche Taycan	150/270kW	79/79/80	199	9.5
		Midsize Cars		
Audi e-tron GT	175kW Asynchron	82/81.83	238	10
Mazda MX-30	81kW AC PMSM	92/98/85	100	5.3
Nissan Leaf 40kW	110kW DCPM	111/123/99	149	8
Porsche 4 Cross	175 and 320kW ACPM	76/76/77	215	10.5
TESLA 3 long range	98/165kW AC-3 phase	131/134/126	358	9.6/11.5[a]
		Large Cars		
Hyundai Ioniq 5 AWD	74/165kW PMSM	98/110/87	256	8.5
TESLA S	247/247kW AC 3-phase	120/124/115	405	8.3/15[a]
		Small Station Wagon		
Chevrolet Bolt EUV	150kW ACPM	115/125/104	247	7.5
KIA EV6 AWD	74/165kW PMSM	105/116/94	274	8.4
		Standard Pickup Truck 2WD		
Ford F-150 BEV 4x2	NA	NA	NA	NA
		Standard Pickup Truck 4WD		
Ford 150 Lightning	358kW AC PMSM	68/76/61	230	10
		Small Sport Utility Vehicles 2WD		
Ford Mustang MACH-E	216kW AC PMSM	101/108/94	314	10.1
Hyundai Kona e	150kW AC PMSM	120/132/108	258	9.5
TELSA Model Y RWD	209kW AC 3-phase	129/140/119	244	4.4/8[c]
Volkswagen ID-4	150kW AC 3-phase	107/116/98	275	7.5
		Small Sport Utility Vehicles 4WD		
Ford AWD Mustang Mach-E	198/198kW AC PMSM	93/99/86	224	8

(*Continued*)

TABLE 14.1 (*Continued*)
Selected EV Models and Fuel Economy/Driving/Charging Parameters

Model	Motor	Fuel Economy (MPGe)	Driving Range (comb city/hwy)	Charge Time (hrs @ 240V)
Mercedes-Benz EQB 300 4M	396V AC1 ASM 1	NA	NA	NA
TESLA Model Y	91/200kW AC	123/129/116	279	9.4
Volkswagen ID.4 AWD Pro	80/150kW AC	101/106/96	251	7.5
Volvo C40	150/150kW AC	87/94/80	226	8
		Standard Sport Utility Vehicles 4WD		
Audi e-tron quattro	141/172kW Asynchron	78/78/77	222	10
BMW iX50 20in wheels	190/230kW EESM	86/86/87	324	12
Rivian R15	162/163kW AC	69/73/65	316	13
TESLA Model X	243/248kW AC	102/107/97	348	8/14[b]

ACPM, Alternating current permanent magnet motor; DCPM, Direct current permanent magnet motor; EESM, Externally excited synchronous machine; PMSM, Permanent magnet synchronous motor.

[a] Range for combined city/highway driving (55% city and 45% highway).

[b] First value is time required with the 48A high-power options; second value is with standard charger.

[c] The first value is time required with the 72A high-power charger option; second value is with standard charger.

TABLE 14.2
Selected Fuel Cell Vehicles

Model (miles)	Fuel Cell Type	Fuel Type	Driving Range
	Compact Cars		
Toyota			
Miral LE	PEFC	Hydrogen	330
Miral Limited	PEM	Hydrogen	357
Miral XLE	PEFC	Hydrogen	402
	Small Sport Utility Vehicles 2WD		
Hyundai			
Nexo	PEM	Hydrogen	354
Nexo Blue	PEM	Hydrogen	380

PEFC, Polymer electrolyte fuel cell; PEM, Proton Exchange Membrane.

The current issue is that hydrogen fuel cell vehicles are still in the process of evolving technology and have several challenges that have to be met. Table 14.2 lists a few of the current fuel cell car models propelled by fuel cells along with the fuel used and their driving ranges.

PLUG-IN HYBRID VEHICLE

Plug-in hybrid electric vehicles (PHEVs) are hybrids that can be charged by plugging them into a charging station or an electrical outlet. Under typical driving conditions plug-in hybrids can store enough electricity from the power grid to significantly reduce their gasoline consumption. There are two basic plug-in hybrid configurations:

- **Series PHEVs, also called Extended-Range Electric Vehicles (EREVs)**—the electric motor on these vehicles is the only power source that turns the wheels; the gasoline engine only generates electricity. Series PHEVs can run solely on electricity until the battery needs to be recharged. The gasoline engine will then generate the electricity needed to power the electric motor. These vehicles may not use any gasoline during short trips.
- **Parallel or Blended PHEVs**—both the engine and electric motor are mechanically connected to the wheels, and both may propel the vehicle. The vehicle may operate using both electricity and gasoline at the same time, using gasoline only, or using electricity only.

Plug-in hybrids also have different battery capacities, allowing some to travel farther on electricity than others; however, driving style, driving conditions, and accessory use like that of EVs can affect not only operation but also range. In pure electric mode of operation, PHEVs have no tailpipe pollutants, although the power plant producing the electricity may emit pollution.

Charging a PHEV's battery typically takes several hours. They can be charged at home or at an increasing number of workplaces or public locations. They can be fueled solely by gasoline either by the driver's choice or by need, like the conventional hybrid. To achieve maximum range or fuel economy, however, they must be charged.

The bottom line: plug-in hybrids use less gasoline and cost less to fuel than conventional hybrids, but they are more expensive to purchase. Table 14.3 lists selected plug-in hybrid vehicles and key parameters.

ETHANOL FLEXIBLE FUEL VEHICLES

Ethanol flexible fuel vehicles (FFVs) are designed to operate on gasoline, E85, or any mixture of the two fuels. The price of ethanol is highly variable from region to region. Note that it is typical in the Midwestern United States and in higher areas. In Table 14.3 fuel and driving range are shown. When the FFV is operated on mixtures of gasoline and E85 such as when alternating between using these fuels, driving range and fuel economy values will likely be somewhere between those listed for two fuels. Note that this depends on the actual percentages of gasoline and E85 in the tank. Table 14.4 lists the models, their fuel, and driving range in miles.

MODEST PROPOSAL #1

In Modest Proposal number 1, I suggest, recommend, advocate, and advise that the number one problem with non-conventional pure electric vehicles-on-wheels is range—it is all about range. This is not rocket science—many drivers like to drive long distances

TABLE 14.3
Selected Plug-In Hybrid Vehicles

Model	Fuel	Range (Rounded to Nearest 10 miles)	Charge Time
	Two-Seater Cars		
Ferrari			
296 GTB	Electricity + Gasoline	350	2.5
SF90 Spider	Electricity + Gasoline	330	2.5
SF90 Stradale	Electricity + Gasoline	330	2.5
	Compact Cars		
BMW			
330e Sedan	Electricity + Gasoline	320	3
350e xDrive Sedan	Electricity + Gasoline	290	3
530e Sedan	Electricity + Gasoline	340	3
Volvo			
S60 T8 AWD	Electricity + Gasoline	510	3
S60 T8 AWD Extended Range	Electricity + Gasoline	530	5
	Midsize Cars		
Audi			
A7 TFSI e quattro	Electricity + Gasoline	410	3
Bentley			
Flying Spur Hybrid	Electricity + Gasoline	430	3
Hyundai			
Ioniq Plug-in Hybrid	Electricity + Gasoline	620	2.2
Mini			
Cooper SE Countryman AII4	Electricity + Gasoline	300	2
Toyota			
Prius Prime	Electricity + Gasoline	640	2
Volvo			
S90 T8 AWD Recharge	Electricity + Gasoline	490	3
S 90 T8 AWD Recharge ext. Range	Electricity + Gasoline	500	5
	Large Cars		
BMW 745e X Drive	Electricity + Gasoline	290	4
Porsche			
Panamera 4 E-Hybrid/ Exec/ST	Electricity + Gasoline	480	3
Panamera 4S E-hybrid/ Exec/ST	Electricity + Gasoline	480	3

(Continued)

TABLE 14.3 (*Continued*)
Selected Plug-In Hybrid Vehicles

Model	Fuel	Range (Rounded to Nearest 10 miles)	Charge Time
	Small Station Wagons		
KIA			
Niro Plug-in Hybrid	Electricity + Gasoline	560	2.2
Volvo			
V-60 T8AWD Recharge	Electricity + Gasoline	530	3
	Minivans 2WD		
Chrysler			
Pacifica Hybrid	Electricity + Gasoline	520	2
	Small Sport Utility Vehicles 2WD		
Ford			
Escape FWD PHEV	Electricity + Gasoline	520	3.3
	Small Sport Utility Vehicles 4WD		
AUDI			
Q5 TFSI e Quattro	Electricity + Gasoline	390	3
Hyundai			
Santa Fe Plug-in Hybrid	Electricity + Gasoline	440	3.4
Tucson Plug-in Hybrid	Electricity + Gasoline	420	1.7
Jeep			
Wrangler 4dr 4xe	Electricity + Gasoline	370	2.4
KIA			
Sorento Plug-in Hybrid	Electricity + Gasoline	460	3.4
LEXUS			
NX 450h Plus AWD	Electricity + Gasoline	550	4.5
Lincoln			
Corsair AWD PHEV	Electricity + Gasoline	430	3.5
Mitsubishi			
Outlander PHEV	Electricity + Gasoline	320	4
Toyota			
RAV 4 Prime 4WD	Electricity + Gasoline	600	4.5
Volvo			
XC60 T8 AWD Recharge	Electricity + Gasoline	500	3
	Standard Sport Utility Vehicles 4WD		
BMW			
X5 xDrive45e	Electricity + Gasoline	400	5
Jeep			

(Continued)

TABLE 14.3 (*Continued*)
Selected Plug-In Hybrid Vehicles

Model	Fuel	Range (Rounded to Nearest 10 miles)	Charge Time
Grand Cherokee 4xe	Electricity + Gasoline	470	3.4
Land Rover			
Range Rover Sport PHEV	Electricity + Gasoline	480	3
Lincoln			
Aviator PHEV AWD	Electricity + Gasoline	460	3.5
Porsche			
Cayenne Turbo S/ Coupe E-Hybrid	Electricity + Gasoline	430	3
Volvo			
XC90 T8 AWD Recharge	Electricity + Gasoline	520	3

TABLE 14.4
Ethanol Flexible Fuel Vehicles Fuel and Driving Range

Model	Fuel	Driving Range
	Two-Seater Cars	
Koenigsegg Automobile AB (mid-engine sports car)		
Jesko	NA	NA
	Standard Pickup Trucks 2WD	
Chevrolet		
Silverado 2WD	Gas/E85	384/288
Ford		
F150 Pickup 2WD FFV	Gas/E85	503/383
GMC		
Sierra 2WD	Gas/E85	384/288
	Standard Pickup Trucks 4WD	
Chevrolet		
Silverado 4WD	Gas/E85	384/288
Silverado Mud Terrain Tires 4WD	Gas/E85	360/264
Ford		
F150 Pickup 4WD FFV A-S10, 3.3L6cyl	Gas/E85	478/358
A-S10, 5.0L, 8cyl	Gas/E85	454/513

(*Continued*)

TABLE 14.4 (*Continued*)
Ethanol Flexible Fuel Vehicles Fuel and Driving Range

Model	Fuel	Driving Range
GMC		
Sierra 4WD	Gas/E85	384/288
Sierra Mud Terrain Tires 4WD	Gas/E85	360/264
	Vans, Passenger Type	
Ford		
Transit T150 Wagon 2WD FFV	Gas/E85	420/296
Transit T150 Wagon 4WD FFV	Gas/E85	395/300
Special Purpose Vehicles 2WD		
Ford		
Transit Connect Van FFV	Gas/E85	395/300
Transit Connect Wagon LWB FFV	Gas/E85	411/284
	Standard Sport Utility Vehicles 4WD	
Ford		
Explorer AWD FFV	Gas/E85	Na/Na
Explorer FFV AWD	Gas/E85	414/283

without delay. Range, of course, is a limiting factor because whether your non-conventional vehicle can be driven 200 or 400 miles without the recharging process, it can't compare with conventional propelled vehicles that can simply pull into a gas station when their tank is low and refill in a matter of a few minutes and then hit the highway again and move on until they need to refill again. You can't do that with a pure electric vehicle.

So, what is the suggestion, recommendation, and advice about powering and extending the range of pure electric vehicles?

Good question.

My suggestions begin with finding a way to reduce battery size into a cylinder shape of 4 feet in length and 6 inches in diameter. At the present time, batteries in electric vehicles are large and occupy almost the entire area of the undercarriage of the electric vehicle. Therefore, batteries of such size are not easily removed from the vehicle and installing a new one is no charm either.

So, what is the point here? The point is if you can go to a standard gas station throughout the United States and Canada and other locations and change out a battery for one that is fully charged, it is like going in and instead of pumping gas just making a quick change out with a new fully charged battery to continue on with their travels. Not only is it about range but also about convenience—being able to change out a battery and travel onward without time lost is priceless, especially for long-distance driving.

Ok, so, explain.

Here's the deal. If we can make the batteries for pure electric vehicles-on-wheels just as easy to change out as it is to fill a conventional vehicle with gasoline, then the market will explode because drivers will be more inclined to purchase pure electric vehicles—it is all about convenience.

Okay, how about the number one factor—range?

If you can change the change out battery when you stop at a gasoline/battery station, range is unlimited. Also, the cylindrical batteries can be inserted piggy-back if properly configured. What this means is that the range of the vehicle can be doubled or increased even more. For example, if a semi-truck piggy backs two on one side of the truck and two more on the other side then mileage and range can be increased, bigtime.

MODEST PROPOSAL #2

Ok, after proposing a cylinder-shaped battery that can fit into any vehicle-on-wheels and stating that these batteries should be easy change-outs at every forthcoming gas/battery change out stations what else needs to be said?

Well, another good question.

Here's the proposal: find a way to recharge vehicle-on-wheels batteries at all times.

At all times?

Yes.

How? Is the vehicle to be plugged into the grid at all times, even when driving on the highway?

No, not exactly. In this proposal, I suggest that researchers find a way in which solar power and other light sources (such as garage lighting) react with a to-be-developed Ultra-Light-Sensitive Solar Cell (aka electricity-producing light cells) combination. In an ideal circumstance (beyond the present wishful thinking) we will develop light cells that not only react to sunlight (daylight) but also react to incandescent or fluorescent lighting that is taken in by the Ultra-Light-Sensitive Cells and converted to electricity to power the vehicle's traction motor.

Well, that sounds good but what do we do when the sun sets and nightfall sets in with darkness and no sunlight?

Another good question.

Ok, when driving anywhere on any road (or other) during night-time with no sunlight the solution to recharging is the vehicle's headlights. If super-sensitive solar light detectors are placed within the headlight compartments with mirrors enclosed, the light from the headlights will be converted to charging the battery.

Now, in this day- and night-time charging process, how long are the new cylinder-shaped, 4-foot-long slide-in and -out batteries increasing their range to at least 10,000 miles before change out must be made?

Another good question.

If we can build (and we can) a vehicle charging system that is ongoing in almost all circumstances via a trickle charging function that charges an electric vehicle-on-wheels battery in order to maintain a flow of electricity through the vehicle's system to the traction motor then customers will flock to the dealerships to purchase one of these vehicles.

The bottom line with electric vehicles-on-wheels is that it is all about range—increase the range and a follow-up purchase of such vehicles may be in the exponential range.

15 Electric Vehicles
The Future

INTRODUCTION

Tooling down the highway with no worries about your vehicle's operation and with no worries at all about the propulsion system getting you from place to place is something we all want. The question is: can electric-powered vehicles-on-wheels accomplish this?

Good question.

Yes, there is a future for electric vehicles. The problem is that the future is developing and is not exactly present. Meaning, technological development and advancement are still needed to make electric vehicles standard—everywhere.

What it all boils down to is whether to purchase an electric vehicle-on-wheels or not—this is consideration that each buyer must confront.

So, what are the considerations involved in deciding to purchase an electric vehicle?

Well, when shopping for an electric car the buyer has to face a specific set of concerns or questions. These questions are addressed in the following but be advised the issues discussed are not the only concerns the informed vehicle purchaser might have.

FOR A FEW MILES MORE

Earlier, it was pointed out that the number one concern potential purchasers of electric vehicle deal with is range. Specifically, the electric vehicle shopper's most important concern is about the range of the vehicle in relation to his or her needs. That is, is the purchaser going to be content with a travel range of 100 to a bit more than 300 miles without stopping to recharge the vehicle's battery? In a non-scientific study, a poll that I conducted questioning more than 100 vehicle commuters (mostly active-duty military men and women) in the Norfolk/Virginia Beach region of Virginia, I first asked each of those polled about the round-trip commute they had to make each work day or work night. Roughly speaking the round-trip in miles turned out to be 30–35 miles on average. My next question to the respondents was, do they think they would be comfortable in purchasing an electric vehicle to make their commute from home to work each workday? Turns out that 72 out of 104 respondents (again, mostly active-duty military) said they did not want pure electric vehicles, but about half this group stated that they would purchase a hybrid and 12 of the respondents were already driving hybrid vehicles. Of the 12 driving hybrids all of them said they were satisfied with their vehicle's operation and the vehicle's range. My next question to the respondents in regard to electric vehicle range was whether they actually

DOI: 10.1201/9781003332992-15

believed that the range touted by the car manufacturers is accurate—if advertised that a vehicle has a 200-mile range before recharging is that necessarily accurate; is it true? Almost all the respondents, including hybrid owners (none at this time owned a pure electric vehicle) stated that they did not believe the manufacturers' mileage range.

Did this last answer surprise me? No, it did not. I actually agreed with the respondents because I had visited various car dealerships in the local area (Tidewater or Hampton Roads, Virginia). And when I was at four different Hampton Roads dealerships that sell electric and/or hybrids (I was tire-kicking and not buying) I always informed the salesperson that I needed to drive either the pure electric or hybrid for a few days to make up my mind about purchasing one.

No problem. I drove several of those vehicles and drove them as far as I could. In each and every case, the manufacturers' statement of driving range was and is wrong. In one particular instance I drove an electric car for 3 days (a test run) and found that the estimated range of at least 200 miles was not accurate. In one case, I was dead on the highway, so to speak, when the battery discharged totally and that was at 140 miles, total.

So, are the vehicle manufacturers and dealers lying about range?

No, not really. The mileage was determined in a controlled manner during the testing phase conducted by manufacturers, official observers, and other note-takers—and the results are nothing more than estimates.

Highway driving by various drivers, vehicle owners, with several driving techniques is definitely different than any other experimental testing function and often does not meet the estimates.

Basically, driving on the roads, highways, and Interstates with assorted drivers is different than controlled laboratory testing.

And this was the result of my investigation. Again, non-scientific but realistic, as far as I am concerned.

FOR A FEW DOLLARS MORE

When I conducted my unscientific polling or survey of various drivers in the Hampton Roads region of Virginia, not only did I hear that range was a concern but also pricing. The average person (as judged by me) was and is looking for a reliable vehicle that is affordable.

Ok, what is affordable—what does that mean? Well, for the folks I talked to during my non-scientific polling I found out that most of the potential electric vehicles buyers were gun-shy, so to speak, about paying $60,000 for any vehicle. Simply stated, next to an electric car's operating range, the sticker price is its most important consideration. Even simpler, the average person does not want to spend or commit to buying a vehicle that is in the 60K range. However, the truth be known, the least expensive electric cars for 2022 have prices in the $30,000 range, though some luxury models sell for a lot more. However, most electric cars are eligible for a one-time $7,500 federal tax credit that effectively cuts the price by that amount—other models like Tesla and Chevrolet Bolt EV are only eligible for a smaller federal credit, which is being phased out. The problem is the wait.

The wait?

Yes. The wait is the amount of time one has to wait for the refund or credit on the purchaser's taxes. The present trend seems to be to purchase used models—reduces total cost for many consumers. Based again on my unscientific study of electric cars, I have found that electric cars are much more affordable in the preowned market.

ROOM AND CARGO SPACE AT A PREMIUM

Aside from wondering about electric cars' range and price many respondents of my unscientific polling stated that room and cargo space were a major concern. Many said that the interior was too cramped for the taller motorists. Also, back seat leg room and headroom were a concern when they transported multiple passengers. The concern I heard most often was about enough space for children to be easily accommodated within and trunk room for grocery bags—many of the respondents who had looked at electric cars (the tire kickers) were not happy with the cargo space the sedans provided.

EASY TO DRIVE OR NOT?

A test drive of any new vehicle-on-wheels is important and is almost always conducted by potential purchasers and is especially accomplished before the purchaser signs a bill of sale. Driving an electric vehicle-on-wheels is a somewhat different experience than driving a conventional vehicle. The first difference the driver will notice when test driving the electric vehicle-on-wheels is the lack of noise—no gas or diesel engine and no exhaust, just quiet. The test drive will also have to learn how to use the accelerator because the electric vehicle delivers full power and that takes some getting used to. Another new driving sensation that the driver of an electric vehicle must get used to is the regenerative braking system in some vehicles—the high-energy braking may make the driver uncomfortable.

Low on Kilowatts

Before purchasing an electric vehicle-on-wheels the purchaser must think about charging options. Earlier it was pointed out that charging the plug-in vehicle revolves around three possibilities (at the present time) and even if the technology advances to the point where battery size is drastically reduced and is capable of constantly being charged by exposure to daylight and artificial lighting the vehicle owner may desire to top off the charge, so to speak, by plugging into a charging source.

So, having said all this it is time for a bit of redundancy to aid in setting the stage for the material that follows. The following is the repeated information about the levels of charging for the plug-in electric vehicle.

- **Level 1 (slow) charging (L1)**—every EV comes with a universally compatible L1 charge cable that plugs into any standard grounded 120-V outlet. The L1 charger power rating tops out a 2.4 kW, restoring about 5–8 miles

per hour charge time, about 40 miles every 8 hours. Many drivers refer to the L1 charge cable as a trickle charger or emergency charger. The L1 charger will not keep up with long commutes or long drives anywhere.
- **Level 2 (fast) charging (L2)**—this charger runs at higher input voltage, 240 V, and is usually a dedicated 240-V circuit in a driveway or garage that can be used with a J-plug connector. This is the most common charging system for residential use and can be found at commercial facilities. These chargers tend to top out at 12 kW, restoring up to 12–25 miles per hour charge, about 100 miles every 8 hours.
- **Level 3 (rapid) charging (L3)**—the fastest EV chargers available; it charges a battery to 80% in 30 minutes (using 480-V circuits), then slows to prevent overheating of the battery. Both CHAdeMO and SAE CCS connectors are used.

Let's first look a bit closer at Level 1 plug-in electric vehicle battery charging. Many EV owners are able to meet their daily driving requirements by simply plugging into Level 1 equipment in the house and charging overnight. This Level 1 charging at home requires no additional cost. Of course this is the case only if a power outlet is provided or a dedicated branch circuit is available at their home or at their parking location.

Okay, great, how much does it cost the customer to recharge his or her plug-in electric vehicle-on-wheels? Keep in mind that the fuel efficiency of an EV may be measured in kilowatt-hours (kWh) per 100 miles. To calculate the cost per mile of an EV, the cost of electricity (in dollars per kWh) and the efficiency of the vehicle (how much electricity is used to travel 100 miles) must be known. The US Department of Energy's Energy Efficiency and Renewable Energy (EERE) in *Alternative Fuels Data Center* (2022) provided at https://www.usa.gov provides the following example about recharging costs:

> If electricity costs ¢10.7 per kWh and the vehicle consumes 27 kWh to travel 100 miles, the cost per mile is about $0.03. Now, if electricity costs ¢10.7 per kilowatt-hour, charging an EV with a 200-mile range (assuming a fully depleted 54 kWh battery) will cost about $6 to reach a full charge.

So, the point is that for EV charging, the stability and planning benefits of household electricity rates offer an attractive alternative compared to traditional types of transportation.

CHARGING STATIONS

Earlier I made the point that as technology advances we should come to the point whereby EV batteries will be made smaller, cylindrical in shape, easy to change out for a fully charged battery or set of batteries. Moreover, these change out batteries will have a range of at least 10,000 miles (under normal conditions and normal driving methods). These 10,000-mile batteries will be trickle charged anytime the vehicle is exposed to daylight and artificial light—at night, the headlights will provide the light needed to charge the batteries while traveling.

This is the pipe dream. Well, was there cure for chicken pox, polio, and did we come up with the theory of evolution by natural selection (I proved this is the case to myself when I studied Darwin's Finches in Galapagos Islands)? Did we discover X-rays? Did we come up with the theory of relativity? Did we come up with the discovery of DNA? Did we discover electricity and how to use it? Have we figured out gravity? How about Copernicus and his theory of Heliocentrism? Did we figure out the uniqueness of fingerprints? Oh, and then there is the biggest discovery of them all—did we discover the wheel?

You might think that the questions just asked are silly or maybe ridiculous and you might say that we do not know completely what we do not know completely about all of these monumental discoveries and you would be correct, of course. But we know enough and for now that is enough. The point is EVs are in the earliest stages of technological advancements. And more profound discoveries on how to improve their performance and range are out there—we simply need to find them.

Remember that every problem has a solution.

At the present time, EVs are climbing in popularity and are being ordered at increasing rates. Customers realize that electric vehicles-on-wheels are much more costly than their gasoline counterparts and are out of reach for many buyers even when the fuel savings are factored in. However, many buyers see and feel the trend that is escalating toward the purchase of EVs and many of these buyers feel that the movement toward clean energy is important and growing.

And that is the bottom line.

ENVIRONMENTAL IMPACT OF HYDROGEN FUEL CELLS

Beyond the expectation that hydrogen leakage from its use in fuel cells could greatly impact the hydrogen cycle and could, when oxidized in the stratosphere, cool the stratosphere and create more clouds, delaying the breakup of the polar vortex at the poles, making the holes in the ozone layer larger and longer lasting, little is understood about how hydrogen leakage would affect the environment. For example, much uncertainty exists over the extent of hydrogen emissions' impact on soil absorption of hydrogen from the atmosphere. This concept or principle is important because if we use extensive quantities of hydrogen for fuel cells, absorption of hydrogen by soils could have a compensatory effect on any possible anthropogenic emissions.

Little is understood about how hydrogen leakage would affect the environment. Again, we do not know what we do not know about the impact of hydrogen cells on the environment.

Glossary[1]

A

AC Generator (or Alternator)—an electric device that produces an electric current that reverses direction many times per second. Sometimes called a synchronous generator.

Adsorption—the adhesion of the molecules of gases, dissolved substances, or liquids to the surface of the solids or liquids with which they are in contact.

Air—the mixture of oxygen, nitrogen, and other gases that, with varying amounts of water vapor, forms the atmosphere of the earth.

Alkaline Fuel Cell (AFC)—a type of hydrogen/oxygen fuel cell in which the electrolyte is concentrated potassium hydroxide (KOH) and the hydroxide ions (OH–) are transported from the cathode to the anode.

Alloy—mixture containing mostly metals. For example, brass is an alloy of copper and zinc. Steel contains iron and other metals but also carbon.

Alternating Current (AC)—a type of current that flows from positive to negative and from negative to positive in the same conductor.

Alternative Fuel—an alternative to gasoline or diesel fuel that is not produced in a conventional way from crude oil. Examples include compressed natural gas (CNG), liquefied natural gas (LNG), ethanol, methanol, and hydrogen.

Ambient Air—the surrounding of a given object or system.

Ambient Temperature—the temperature of the surrounding medium, usually used to refer to the temperature of the air in which a structure is situated or a device operates.

Anion—a negatively charged ion; an ion that is attracted to the anode.

Anode—the electrode at which oxidation (a loss of electrons) takes place. For fuel cells and other galvanic cells, the anode is the negative terminal; for electrolytic cells (where electrolysis occurs), the anode is the positive terminal.

Atmospheric Pressure—the force exerted by the movement of air in the atmosphere, usually measured in units of force per area. For fuel cells, atmospheric pressure is usually used to describe a system where the only pressure action on the system is from the atmosphere; no external pressure is applied.

Atom—the smallest physical unit of a chemical element that can still retain all the physical and chemical properties of that element. Atoms combine to form molecule, and they themselves contain several kinds of smaller particles. An atom has a dense central core (the nucleus) consisting of positively charged particles (protons) and uncharged particles (neutrons).

[1] Based on *U.S. Energy Efficiency & Renewable Energy—Hydrogen and Fuel Cell Technologies Office* (2022). Washington, DC: US Department of Energy; USDOE (2022) *Vehicle Technologies Program Westie Glossary*. Washington, DC: U.S. Department of Energy.

Negatively charged particles (electrons) are scattered in a relatively large space around this nucleus and move about it in orbital patterns at extremely high speeds. An atom contains the same number of protons as electrons and thus is electrically neutral (uncharged) and stable under most conditions.

B

Battery—an energy storage device that produces electricity by means of chemical actin. It consists of one or more electric cells, each of which has all the chemicals and parts needed to produce an electric current.

Bipolar Plates—The conductive plate in a fuel cell stack that acts as an anode for one cell and a cathode for the adjacent cell. The plate may be made of metal or a conductive polymer (which may be a carbon-filled composite). The plate usually incorporates flow channels for the fluid beds and many also contain conduits for heat transfer.

British Thermal Unit (Btu)—the mean British thermal unit is 1/180 of the heat required to raise the temperature of one pound (1 lb) of water from 32°F to 212°F at a constant atmospheric pressure. The Btu is equal to the quantity of heat required to raise one pound (1 lb) of water by 1°F.

C

Carbon (C)—an atom and primary constituent of hydrocarbon fuels. Carbon is routinely left as a black deposit on engine parts, such as pistons, rings, and valves, by the combustion of fuel.

Carbon Dioxide (CO_2)—a colorless, odorless, noncombustible gas that is slightly more than 1.5 times as dense as air and becomes a solid (dry ice) below –78.5°C. It is present in the atmosphere as a result of the decay of organic material and the respiration of living organisms. It is produced by the burning of wood, coal, coke, oil, natural gas, or other fuels containing carbon.

Carbon Monoxide (CO)—a colorless, odorless, tasteless, poisonous gas that results from incomplete combustion of carbon and oxygen.

Catalyst—a chemical substance that increases the rate of reaction without being consumed; after the reaction, it can potentially be recovered from the reaction mixture and is chemically unchanged. The catalyst lowers the activation energy required, allowing the reaction to proceed more quickly or at a lower temperature. In fuel cell, the catalyst facilitates the reaction of oxygen and hydrogen. It is usually made of platinum powder very thinly coated onto carbon paper or cloth. The catalyst is rough and porous so the maximum surface area of the platinum can be exposed to the hydrogen or oxygen. The platinum-coated side of the catalyst faces the membrane in the fuel cell.

Catalyst Poisoning—the process of impurities binding to a fuel cell's catalyst, lowering the catalyst's ability to facilitate the desired chemical reaction.

Cathode—the electrode at which reduction (a gain of electrons) occurs. For fuel cells and other galvanic cells, the cathode is the positive terminal, for electrolytic cells (where electrolysis occurs), the cathode is the negative terminal.

Cation—a positively charged ion.
Celsius—the metric temperature scale and unit of temperature (°C). Named after Swedish astronomer Anders Celsius (1701–1744), even though the thermometer first advocated by him in 1743 had 100° as the freezing point of water and 0° as the boiling point, the reverse is the modern Celsius scale. Also called the Centigrade scale (Latin for "hundred degrees").
Centimeter (cm)—a metric unit of linear measure. One centimeter equals about 0.4 inches, and one inch equals about 2.5 centimeters. One foot is equal to approximately 30 centimeters.
Combustion—the burning fire produced by the proper combination of fuel, heat, and oxygen. In the engine, the rapid burning of the air-fuel mixture that occurs in the combustion chamber.
Combustion Chamber—in an internal combustion engine, the space between the top of the piston and the cylinder head in which the air-fuel mixture is burned.
Composite—material created by combining materials differing in composition or form on a macroscale to obtain specific characteristics and properties. The constituents retain their identity; they can be physically identified, and they exhibit an interface among one another.
Compressed Hydrogen Gas (CHG)—hydrogen gas compressed to a high pressure and stored at ambient temperature.
Compressed Natural Gas (CNG)—mixtures of hydrocarbon gases and vapors consisting principally of methane in gaseous form that has been compressed.
Compressor—a device used for increasing the pressure and density of gas.
Cryogenic Liquefaction—the process through which gases such as nitrogen, hydrogen, helium, and natural gas are liquefied under pressure at very low temperatures.
Current Collector—the conductive material in a fuel cell that collects electrons (on the anode side) or disburses electrons (on the cathode side). Current collectors are microporous (to allow fluid to flow through them) and lie in between the catalyst/electrolyte surfaces and the bipolar plates.

D

Density—the amount of mass in a unit volume. Density varies with temperature and pressure.
Direct Methanol Fuel Cell (DMFC)—a type of fuel cell in which the fuel is methanol (CH_3OH) in gaseous or liquid form. The methanol is oxidized directly at the anode instead of first being reformed to produce hydrogen. The electrolyte is typically a PEM.
Dispersion—the spatial property of being scattered over an area or volume.

E

Electrode—a conductor through which electrons enter or leave an electrolyte. Batteries and fuel cells have a negative electrode (the anode) and a positive electrode (the cathode).

Electrolysis—a process that uses electricity, passing through an electrolytic solution or other appropriate medium, to cause a reaction that breaks chemical bonds (e.g., electrolysis of water to produce hydrogen and oxygen).

Electrolyte—a substance that conducts charged ions from one electrode to the other in a fuel cell, battery, or electrolyzer.

Electron—a stable atomic particle that has a negative charge; the flow of electrons through a substance constitutes electricity.

Emission Standards—regulatory standards that govern the amount of a given pollutant that can be discharged into the air from a given source.

Endothermic—a chemical reaction that absorbs or requires energy (usually in the form of heat).

Energy—the quantity of work a system or substance is capable of doing, usually measured in British thermal units (Btu) or Joules (J).

Energy Content—amount of energy for a given weight of fuel.

Energy Density—amount of potential energy in a given measurement of fuel.

Engine—a machine that converts heat energy into mechanical energy.

Ethanol (CH_3CH_2OH)—an alcohol containing two carbon atoms. Ethanol is a clear, colorless liquid and is the same alcohol found in beer, wine, and whiskey. Ethanol can be produced from cellulosic materials or by fermenting a sugar solution with yeast.

Exhaust Emissions—materials emitted into the atmosphere through any opening downstream of the exhaust ports of an engine, including water, particulates, and pollutants.

Exothermic—a chemical reaction that gives off heat.

F

Fahrenheit—a temperature scale and unit of temperature (°F) named for German physicist Gabriel Daniel Fahrenheit (1686–1736), who was the first to use mercury as a thermometric fluid in 1714.

Flammability Limits—the flammability range a gas is defined in terms of its lower flammability limit (LFL) and its upper flammability limit (UFL). Between the two limits is the flammable range in which gas and air are in the right proportions to burn when ignited. Below the lower flammability limit, there is not enough fuel to burn. Above the higher flammability limit, there is not enough air to support combustion.

Flashpoint—the lowest temperature under very specific conditions at which a substance will begin to burn.

Flexible Fuel Vehicle—a vehicle that can operate on a wide range of fuel blends (e.g., blends of gasoline and alcohol) that can be put in the same fuel tank.

Fuel—a material used to create heat or power through conversion in such processes as combustion or electrochemistry.

Fuel Cell—a device that produces electricity through an electrochemical process, usually from hydrogen and oxygen.

Fuel Cell Poisoning—the lowering of a fuel cell's efficiency due to impurities in the fuel binding to the catalyst.

Fuel Cell Stack—individual fuel cells connected in series. Fuel cells are stacked to increase voltage—remember, voltage is pressure and does not flow in an electrical circuit, current flows.

Fuel Processor—device used to generate hydrogen from fuels such as natural gas, propane, gasoline, methanol, and ethanol for use in fuel cells.

G

Gas—fuel gas such as natural gas, undiluted liquefied petroleum gases (vapor phase only), liquefied petroleum gas-air mixtures, or mixtures of these gases.

- **Natural Gas**—mixtures of hydrocarbon gases and vapors consisting principally of methane (CH_4) in gaseous form.
- **Liquefied Petroleum Gases (LPG)**—any material composed predominantly of any of the following hydrocarbons or mixtures of them: propane, propylene, butanes (normal butane or isobutane), and butylene.
- **Liquefied Petroleum Gas-Air Mixture**—liquefied petroleum gases distributed at relatively low pressures and normal atmospheric temperatures that have been diluted with air to produce desired heating value and utilization characteristics.

Gas Diffusion—mixing of two gases caused by random molecular motions. Gases diffuse very quickly, liquids diffuse much more slowly, and solids diffuse at very slow (but often measurable) rates. Molecular collisions make diffusion slower in liquids and solids.

Graphite—mineral consisting of a form of carbon that is soft, black, and lustrous and has a greasy feeling. Graphite is used in pencils, crucibles, lubricants, paints, and polishes.

Gravimetric Energy Density—potential energy in a given weight of fuel.

Greenhouse Effect—warming of the earth's atmosphere due to gases in the atmosphere that allow solar radiation (visible, ultraviolet) to reach the earth's atmosphere but do not allow the emitted infrared radiation to pass back out of the earth's atmosphere.

Greenhouse Gas (GHG)—gases in the earth's atmosphere that contribute to the greenhouse effect.

H

Heat Exchanger—device (e.g., a radiator) that is designed to transfer heat from the hot coolant that flows through it to the air blown through it by the fan.

Heating Value (TOTAL)—the number of British thermal units (Btu) produced by the combustion of one cubic foot of gas at constant pressure when the products of combustion are cooled to the initial temperature of the gas and air, when the water vapor formed during combustion is condensed, and when all the necessary corrections have been applied.

- **Lower (LHV)**—the value of the heat of combustion of a fuel measured by allowing all products of combustion to remain in the gaseous state. This method does not consider the heat energy put into the vaporization of water (heat of vaporization).
- **Higher (HHV)**—the value of the heat of combustion of a fuel measured by reducing all of the products of combustion back to their original temperature and condensing all water vapor formed by combustion. This value considers the heat of vaporization of water.

Hybrid Electric Vehicle (HEV)—a vehicle combining a battery-powered electric motor with a traditional internal combustion engine. The vehicle can run on either the battery or the engine or both simultaneously, depending on the performance objectives for the vehicle.

Hydrides—chemical compounds formed when hydrogen gas reacts with metals. Used for storing hydrogen gas.

Hydrocarbon (HC)—an organic compound containing carbon and hydrogen, usually derived from fossil fuels, such as petroleum, natural gas, and coal.

Hydrogen (H_2)—hydrogen (H) is the most abundant element in the universe, but it is generally bonded to another element. Hydrogen gas (H_2) is a diatomic gas composed of two hydrogen atoms and is colorless and odorless. Hydrogen is flammable when mixed with oxygen over a wide range of concentrations.

Hydrogen-Rich Fuel—a fuel that contains a significant amount of hydrogen, such as gasoline, diesel fuel, methanol (CH_3OH), ethanol (CH_3CH_2OH), natural gas, and coal.

I

Impurities—undesirable foreign material(s) in a pure substance or mixture.

Internal Combustion Engine (ICE)—an engine that converts the energy contained in a fuel inside the engine into motion by combusting the fuel. Combustion engines use the pressure created by the expansion of combustion to produce gases to do mechanical work.

Ion—atom or molecule that carries a positive or negative charge because of the loss or gain of electrons.

K

Kilogram (kg)—metric unit of width or mass equal to approximately 2.2 lb. Related units are the milligram (mg) at 1,000,000 per kg and the metric ton at 1,000 kg.

Kilowatt (kW)—a unit of power equal to about 1.34 horsepower or 1,000 watts.

L

Liquefied Hydrogen (LH_2)—hydrogen in the liquid form. Hydrogen can exist in a liquid state but only at extremely cold temperatures. Liquid hydrogen typically is to be stored at −253°C (−463°F). The temperature requirements for liquid hydrogen storage necessitate expending energy to compress and chill the hydrogen into its liquid state.

Liquefied Natural Gas (LNG)—natural gas in liquid form. Natural gas is a liquid at −162°C (−259°F) at ambient pressure.

Liquefied Petroleum Gas (LPG)—any material that consists predominantly of any of the following hydrocarbons or mixtures of hydrocarbons: propane, propylene, normal butane, isobutylene, and butylene. LPG is usually stored under pressure to maintain the mixture in the liquid state.

Liquid—a substance, unlike a solid, flows readily but, unlike a gas, does not tend to expand indefinitely.

M

Mechanical Energy—energy in mechanical form.

Megawatt (MW)—a unit of power equal to one million watts or 1,000 kilowatts.

Membrane—the separating layer in a fuel cell that acts as an electrolyte (an ion exchanger) as well as a barrier film separating the gases in the anode and cathode compartments of the fuel cell.

Meter (m)—basic metric unit of length equal to 3.28 feet, 1.09 yards, or 33.37 inches. Related units are the decimeter (dm) at 10 per meter, the centimeter (cm) at 100 per meter, the millimeter (mm) at 1,000 per meter, and the kilometer (km) at 1,000 meters.

Methanol (CH_3OH)—an alcohol containing one carbon atom. It has been used, together with some of the higher alcohols, as high-octane gasoline component and is a useful automotive fuel.

Miles Per Gallon Equivalent (MPGE)—energy content equivalent to that of a gallon of gasoline (114,320 Btu).

Millimeter (mm)—metric unit of length equal to 0.04 inches. There are 25 millimeters in an inch and 1,000 millimeters in a meter.

Milliwatt (mW)—a unit of power equal to one-thousandth of a watt.

Molten Carbonate Fuel Cell (MCFC)—a type of fuel that contains a molted carbonate electrolyte. Carbonate ions (CO_{3-2}) are transported from the cathode to the anode. Operating temperatures are typically near 650°C.

N

Nafion®—sulfonic acid in a solid polymer form that is usually the electrolyte of PEM fuel cells.

Natural Gas—a naturally occurring gaseous mixture of simple hydrocarbon components (primarily methane) used as a fuel.

Nitrogen (N_2)—a diatomic (consisting of two atoms) colorless, tasteless, odorless gas that constitutes 78% of the atmosphere by volume.

Nitrogen Oxides (NO_x)—any chemical compound of nitrogen and oxygen. Nitrogen oxides result from high temperature and pressure in the combustion chambers of automobile engines and other power plants during the combustion process. When combined with hydrocarbons in the presence of sunlight, nitrogen oxides form smog. Nitrogen oxides are basic air pollutants; automotive exhaust emission levels of nitrogen oxides are regulated by law.

O

Oxidant—a chemical, such as oxygen, that consumes electrons in an electrochemical reaction.

Oxidation—loss of one or more electrons by an atom, molecule, or ion.

Oxygen (O_2)—a diatomic colorless, tasteless, odorless gas that makes up about 21% of air.

P

Partial Oxidation—fuel reforming reaction where it is oxidized partially to carbon monoxide and hydrogen rather than fully oxidized to carbon dioxide and water. This is accomplished by injecting air with the fuel stream prior to the reformer. The advantage of partial oxidation over steam reforming of the fuel is that it is an exothermic reaction rather than an endothermic reaction and therefore generates its own heat.

Pascal (Pa)—the Pascal is the International System of Units (SI)—derived unit of pressure or stress. It is a measure pf perpendicular force per unit area. It is equivalent to one newton per square meter. A megapascal equals 1,00,000 Pascals.

Permeability—ability of a membrane or other material to permit a substance to pass through it.

Phosphoric Acid Fuel Cell (PAFC)—a type of fuel cell in which the electrolyte consists of concentrated phosphoric acid (H_3PO_4). Protons (H^+) are transported from the anode to the cathode. The operating temperature range is generally 160°C–220°C.

Polymer—natural or synthetic compound composed of repeated linings of simple molecules.

Polymer Electrolyte Membrane (PEM)—a fuel cell incorporating a solid polymer membrane used as its electrolyte. Protons (H+) are transported from the anode to the cathode. The operating temperature range is generally 60°C–100°C.

Polymer Electrolyte Membrane and Fuel Cell (PEMFC or PEFC)—a type of acid-based fuel cell in which the transport of protons (H_2) from the anode to the cathode is through a solid, aqueous membrane impregnated with an appropriate acid. The electrolyte is called a polymer electrolyte membrane (PEM). The fuel cells typically run at low temperatures (<100°C).

Proton—a subatomic particle in the nucleus of an atom that carries a positive electric charge and is not movable by electrical means.

R

Reactant—a chemical substance that is present at the start of a chemical reaction.

Reactor—device or process vessel in which chemical reactions (e.g., catalysis in fuel cells) take place.

Reformate—hydrocarbon fuel that has been processed into hydrogen and other products for use in fuel cells.

Reformer—device used to generate hydrogen from fuels such as natural gas, propane, gasoline, and ethanol for use in fuel cells.

Reforming—a chemical process in which hydrogen-containing fuels react with steam, oxygen, or both to produce a hydrogen-rich gas stream.

Reformulated Gasoline—gasoline that is bled so that, on average, it reduces volatile organic compounds and toxic emissions significantly relative to conventional gasolines.

Regenerative Fuel Cell—a fuel cell that produces electricity from hydrogen and oxygen and can use electricity from solar power or some other source to divide the excess water into oxygen and hydrogen fuel to be reused by the fuel cell.

Renewable Energy—a form of energy that is never exhausted because it is renewed by nature (within short times scales; e.g., wind, solar radiation, hydropower).

S

Solid Oxide Fuel Cell (SOFC)—a type of fuel in which the electrolyte is a solid nonporous metal oxide, typically zirconium oxide (ZrO_2) treated with Y_2O_3, and O^{-2} is transported from the cathode to the anode. Any CO in the reformate gas is oxidized to CO_2 at the anode. Temperatures of operation are typically 800°C–1,000°C.

Sorbent—material that sorbs another (i.e., has the tendency to take it up either by adsorption or by absorption).

Sorption—process by which one substance takes up or holds another.

Steam Reforming—the process for reacting a hydrocarbon fuel, such as natural gas, with steam to produce hydrogen as a product. This is a common method for bulk hydrogen generation.

T

Technology Validation—confirming that technical targets for a given technology have been met.

Temperature—a measure of thermal content.

Turbine—machine for generating rotary mechanical power from the energy in a stream of fluid. The energy, originally in the form of head or pressure energy, is converted to velocity energy by passing through a system of stationary and moving blades in the turbine.

Turbocharger—a device used for increasing the pressure and density of a fluid entering a fuel cell power plant using a compressor driven by a turbine that extracts energy from the exhaust gas.

Turbocompressor—machine for compressing air or other fluids (reactant if suppled to a fuel cell system) in order to increase the reactant pressure and concentration.

V

Volumetric Energy Density—potential energy in a given volume of fuel.

W

Water (H_2O)—a colorless, transparent, odorless, tasteless liquid compound of hydrogen and oxygen. The liquid form of steam and ice. Fresh water at atmospheric pressure is used as a standard for describing the relative density of liquids, the standard for liquid capacity, and the standard for fluid flow. The melting and boiling points of water are the basis for the Celsius temperature system. Water is the only by-product of the combination of hydrogen and oxygen and is produced during the burning of any hydrocarbon. Water is the only substance that expands on freezing as well as by heating and has a maximum density at 4°C.

Watt (W)—a unit of power equal to one Joule of work performed per second; 746 watts is the equivalent of one horsepower. The watt is named for James Watt, Scottish engineer (1736–1819) and pioneer in steam engine design.

Wt.%—the term wt.% (abbreviation for weight percent) is widely used in hydrogen storage research to denote the amount of hydrogen stored on a weight basis, and the term mass % is also occasionally used. The term can be used for materials that store hydrogen or for the entire storage system (e.g., material or compressed/liquid hydrogen as well as the tank and other equipment required to contain the hydrogen such as insulation, valves, regulators, etc.). For example, 6 wt.% on a system basis means that 6% of the entire system by weight is hydrogen. On a material basis, the wt.% is the mass of hydrogen divided by the mass of material plus hydrogen.

Index

Note: **Bold** page numbers refer to tables and *italic* page numbers refer to figures.

Printed in the United States
by Baker & Taylor Publisher Services